Pedigree Polytopes

Alan Turing - Celebrating the life of a genius

Tirukkattuppalli Subramanyam Arthanari

Pedigree Polytopes

New Insights on Computational Complexity
of Combinatorial Optimisation Problems

 Springer

Tirukkattuppalli Subramanyam Arthanari
Visitor, Information Systems
and Operations Management
University of Auckland
Auckland, New Zealand

ISBN 978-981-19-9954-3   ISBN 978-981-19-9952-9 (eBook)
https://doi.org/10.1007/978-981-19-9952-9

This Springer imprint is published by the registered company Springer Nature Singapore Pte Ltd.
The registered company address is: 152 Beach Road, #21-01/04 Gateway East, Singapore 189721,
Singapore

A peep into the impossible doesn't demand much
except the courage to face the unexpected.

— *A twist on Arthur C. Clarke's Second Law*

I fondly dedicate this work to
*my wife **Jaya**,*
for her sacrifices, patience and kindness
through over four decades,
accepting me as I am,
without wasting time correcting me,
however, she helped me in correcting this
manuscript.

Preface

My journey as a researcher started in the second year of my master's degree in Statistics with a speciality in operations research at the Indian Statistical Institute (ISI), Calcutta (now known as Kolkata). The year was 1966.

Next year I completed my dissertation for my postgraduate Diploma in Operations Research entitled "On Compound Simple Games", an expository essay. It was at this time Prof. C. R. Rao, insisted that as operations research students we need to get applied EXPERIENCE, so I should forget Game Theory for now and go and solve some real-world problems in industries and businesses. From these applied research projects, my understanding is that some practical problems, apart from being important in value creation, are theoretically challenging as well and therefore worth pursuing as research projects.

One such problem that attracted me as a student was the travelling salesman problem (*TSP*), which seeks a solution to "Given a list of cities and the distances between each pair of cities, what is the shortest possible route that visits each city exactly once and returns to the origin city?"[1] At that time, I was not informed that Prof. P. C. Mahalanobis, the founder of ISI, had worked on *TSP* and provided an estimate of the optimal tour length.

This is the right place to acknowledge, Padma Vibhushan, Prof. P. C. Mahalanobis, FRS, and all my mentors who inspired me, to appreciate learning and logic and above all their applications to compassionate living. Important among them is Centenarian, Padma Vibhushan, Prof. C. R. Rao, FRS.

Problems involving permutations, like machine sequencing and scheduling problems were the research topics for my doctoral thesis [9]. By the time I was completing my research, it was known these problems are difficult combinatorial optimisation problems. However, I have identified polynomial time algorithms for [1] some special cases of the three-machine flow shop problem, [2] the single-machine sequencing problem with intermittent job arrivals to minimise the number of late jobs and [3] the open shop problem to minimise make-span in a batch processing environment. But, the challenges of other sequencing problems and grouping problems in Numerical

---

[1] https://en.wikipedia.org/wiki/Travelling salesman problem.

Taxonomy that I had considered in my doctoral research remained difficult, and I could only design branch-and-bound algorithms for them.

In 1982 I gave a new 0/1 integer formulation of the symmetric travelling salesman problem (STSP). Using an IBM 360 machine, while solving the LP relaxation of this formulation, I found that 98% of the problems solved gave integer optimal solutions. The number of problems (around 100) as well as the size (maximum 20) was small though. I presented it at the XIth International Symposium on Mathematical Programming, which was held at the University of Bonn, Germany.

After another decade or more, I was able to renew my work on the new formulation with my doctoral student, Ms. Usha.[2]

Subsequently, after migrating to New Zealand on 29 June 1996, I had the opportunity to work again on the new formulation, as I was not employed and so had the time I needed to immerse myself in the problem. This resulted in defining the Pedigrees and Pedigree Polytopes. A major event was the two-day workshop in October 1998, organised by Prof. Andy Philpott at the Department of Engineering Science, the University of Auckland, New Zealand. I presented my results on pedigree polytopes, which I entitled, "On Whole Number Solutions". I gave a necessary and sufficient condition for membership in pedigree polytopes.

When Ms. Laleh Haerian Ardekani joined me as a doctoral student, she was able to further explore the multistage insertion formulation (the new formulation mentioned earlier) of STSP.[3]

Now in this book, I bring together in one place my results on pedigree polytopes published in journals Discrete Mathematics, Discrete Applied Mathematics, and Discrete Optimisation and elsewhere as book chapters or journal articles. Most importantly, Chaps. 5 and 6 of this book present new results from my research providing complete proof of efficiently checking membership in pedigree polytopes. Far-reaching implications of this new research, in solving combinatorial optimisation problems in general and new insights into the computational complexity of solving such problems, are the major contributions of this book.

Incidentally, since linear optimisation over the pedigree polytope, $conv(P_n)$ is equivalent to solving the symmetric travelling salesman problem with n cities, this book establishes the theorem $\mathsf{NP} = \mathsf{P}$; where loosely explained, P is the set of relatively easy problems, and NP is the set that includes seemingly difficult to solve hard problems.

So we have a piece of unexpected good news, that is, the hard problems have relatively easy solutions if we observe from this new perspective. This means plenty of

---

[2] Professor Usha Mohan completed her doctoral thesis, "Symmetric Travelling Salesman Problem: Some New Insights", for her Ph.D. degree from the Indian Statistical Institute.

[3] This resulted in Dr. Laleh Ardekani's doctoral thesis, "New Insights on the Multistage Insertion Formulation of the Traveling Salesman Problem-Polytopes, Experiments, and Algorithm", securing her Ph.D. degree from the University of Auckland, Auckland, New Zealand.

opportunities for researchers in computer science, operations research and management science to explore novel solutions to difficult combinatorial optimisation problems. From an economic point of view this implies adding value through solving efficiently real-world problems, that are routinely faced in practice, which seek solutions that are whole numbers.

Auckland, New Zealand   Tirukkattuppalli Subramanyam Arthanari
December 2022

Acknowledgments

I could not have done this research and completed this book project without the constant support, encouragement, and care from my family. They have fondly enquired about the progress of this book all along and did their part in keeping me going ahead enthusiastically with the challenge.

I missed many good opportunities to spend quality time with my grandchildren as I was busy with this project, and I hope they lovingly forgive me for this. I thank my whanau (extended family) with gratitude for their kindness and support.

My friends, Dr. N. D. Prabhakar (formerly Bell Labs Visiting Fellow, Lucent Technologies, USA) and Prof. N. R. Achuthan (formerly with the Department of Mathematics and Statistics, the Curtin University of Technology, Australia), have taken the trouble to read and comment on my writing, encouraging and supporting me for many years during this project. I thank them profusely.

I take this opportunity to thank all my co-authors for their time and effort to collaborate with me, especially Prof. Usha Mohan, Department of Management Studies, IIT Madras, Chennai, India (who was my doctoral student at ISI, Bengaluru), and Dr. Laleh Haerian Ardekani, Senior Manager Application Delivery, chez Mackenzie Investments, Burlington, Ontario, (who was my doctoral student at the University of Auckland), Prof. Matthias Ehrgott, Department of Management Science Lancaster University, UK (who was also Laleh's co-supervisor), and Mr. Kun Qian, Senior Product Manager at Amazon Web Services (AWS), Seattle, USA (who was a post-graduate student at the Department of ISOM, Auckland, New Zealand).

This page would be incomplete if I do not thankfully remember my colleagues at ISOM, especially Prof. Michael Myers, who encouraged me with his support and friendly advice, Ms. Elviera Cowan, who always acted promptly with a willing smile; Mr. Michael Yang and Mr. Shohil Kishore who delivered end-user IT support with a human touch.

No words can fully justify my gratitude for my erstwhile colleague at the Department of ISOM, Dr. Elke Wolfe (presently CEO, alphaloba GmbH, Cologne, Germany), for her friendship and crucial role in finding suitable researchers in the field to review this work.

I'd like to express my heartfelt thanks to the entire Springer Nature production team, especially Ms. Lucie Bartonek, Commissioning Editor, Mr. Karthikeyan Krishnan, and Ms. Emily Wong, for their undivided attention and care in this production.

I thank the anonymous reviewers of the book for their meticulous reviews and valuable suggestions that have resulted in this improved version of the book.

I have benefitted from the research of many a stalwart in the fields of Computer Science, Operations Research and Mathematics and they are too many to name, this is my opportunity to show my gratitude to them all.

Finally, I thank in advance every one of the readers of this book, who will help me further enhance the value of this publication through their appreciation, comments, critiques, and suggestions.

Epigraph

Mathematics as seen by the poet

The enchantment of rhythm is obviously felt in music,
the rhythm which is inherent in the notes and their grouping.
It is the magic of mathematics,
this rhythm which is in the heart of all creation,
which moves in the atom and in its different measures fashions gold and lead,
the rose and the thorn, the sun and the planets,
the variety and vicissitudes of man's history.
These are the dance steps of numbers in the arena of time and space,
which weave the *maya* of appearance,
the incessant flow of changes
that ever is and is not.
What we know as intellectual truth,
is that also not a perfect rhythm of the relationship of facts
that produce a sense of convincingness to a person
who somehow feels that he knows the truth?
We believe any fact to be true because of a harmony,
a rhythm in reason,
the process of which analysable by the logic of mathematics.

—Nobel Laureate Rabindranath Tagore (1861–1941)[4]

Source: Sankhya, the Indian Journal of Statistics is the official publication of the Indian Statistical Institute., Year: 1935, Volume: 2, Part: 1, Page: 1.

---

[4] The formatting is mine.

Contents

Abbreviations

Acronyms

ATSP	Asymmetric Travelling Salesman Problem
Concorde	A program for solving the travelling salesman problem
COP	Combinatorial Optimisation Problem
CPLEX	IBM ILOG CPLEX is a software that solves LP and related problems
DFJ	Dantzig, Fulkerson, and Johnson formulation
FAT	Forbidden Arc Transportation problem
FFF	Frozen Flow Finding problem/algorithm
GUROBI	A mathematical programming software package
LD	Lagrangian Dual
LP	Linear Programming
MI	Multistage Insertion
NP	Non-deterministic polynomial
NP-complete	Non-deterministic polynomial complete
$P_{MI}(n)$	The *MI*-relaxation polytope for n
P	Polynomial
SEP	Subtour Elimination Polytope
STSP	Symmetric Travelling Salesman Problem
TSP	Travelling Salesman Problem
TSPLIB	Is a library of sample instances for the *TSP* and related problems

Symbols

$\delta(v)$	The set of neighbours of a vertex v or edges incident at v
μ_P	Fixed weight associated with a rigid pedigree P in R_k for a k
τ_n	The number of coordinates of $X \in P_n$
$\mathcal{A}(N_k)$	The arc set of network N_k

c	Denotes an effectiveness measure, $c : \mathcal{F} \to \mathbb{R}$
$c.v$	A 0/1 vector giving a subset of the ground set, called a characteristic vector
$conv(F)$	The convex hull of elements in F
$\mathbb{E}$	The ground set
$E(S)$	The set of edges in K_n with both ends in the set of nodes S
$\mathcal{F}$	A nonempty collection of subsets of $\mathbb{E}$, called conbinatorial objects of interest
G_R	Graph of Rigidity for a given pair of pedigrees
K_n	The complete graph on n vertices
$\mathbb{B}$	The binary set, $\{0, 1\}$
N_k	The layered network for k
$\mathcal{P}_n$	The set of pedigrees for n
P_n	The set of characteristic vectors of pedigrees in $\mathcal{P}_n$
$\mathbb{Q}$	Set of rationals
$\mathbb{R}$	Set of reals
R_k	The set of rigid pedigrees for k
$\mathcal{V}(N_k)$	The node set of network N_k
$\mathbb{Z}$	Set of integers
2^S	The power set of a set S

List of Figures

List of Tables

Chapter 1
Prologue

1.1 Prelude to a New Beginning

Who would have thought that one day we have to prove to robots, several times a day, that each of us has a human pedigree and assert "I am not a robot"?

Perhaps Alan Turing. We have come a long way since Turing defined a deceptively simple *a-machine*, which so efficiently abstracted computation and computability.

Donald Knuth in his Preface to The Art of Computer Programming, Volume 4A, Combinatorial Algorithms [106], observes, "Many combinatorial questions that I once thought would never be answered during my lifetime have now been resolved, and those breakthroughs have been due mainly to improvements in algorithms rather than to improvements in processor speeds."

However, problems that can be efficiently solved using linear/convex programming, turn out to be very difficult, when we require the solutions to satisfy the additional requirement that the solutions be integers.

This book is devoted to understanding such challenging problems from a new perspective. To further scope the exposition, we will mainly consider a class of problems that go by the name *Combinatorial optimisation problems (COP).*[1]

1.1.1 What Are Combinatorial Optimisation Problems?

Definition 1.1 (*COP*) Let $\mathbb{E}$ be a finite set called *ground set*. Let $\mathcal{F}$ denote a nonempty collection of subsets of $\mathbb{E}$, called *combinatorial objects of interest*; $\mathcal{F} \subset 2^{\mathbb{E}}$. Let $c : \mathcal{F} \to \mathbb{R}$, denote an *effectiveness measure*. Generically,

---

[1] Notations and terms used in this chapter are available in Chap. 2.

a combinatorial optimisation problem (COP) can be defined as Find $X^* \in \mathcal{F}$ such that

$$c(X^*) = \text{Optimum}_{X \in \mathcal{F}} \, c(X) \tag{1.1}$$

where, $optimum$ stands for $maximum$ or $minimum$ depending on the effectiveness measure, c. Thus, $(\mathbb{E}, \mathcal{F}, c)$ specifies a COP. ♣

We have a variety of combinatorial optimisation problems studied and solved in the fields of graph theory, operations research, and computer science [108].

Example 1.1 Given a directed graph, $G = (V, A)$ with arc lengths $d : A \to \mathbb{R}$, finding a path between two nodes s, t that minimises the sum of arc lengths along the path is called the shortest path problem. This problem can be cast as a COP given by $\mathbb{E} = A$, $\mathcal{F} = \{P \mid P \text{ is an s-t path in G}\}$, and $c(P) = \sum_{a \in P} d(a)$, $P \in \mathcal{F}$.

Example 1.2 Consider an editor of a journal who has a set of n articles of importance to be published in her next issue of the journal. The publisher has set the maximum number of pages for the issue as W. Given the value for the readers of the journal, and the number of pages required for the article j, $v_j > 0$, and $p_j > 0$ respectively for $j \in [n]$, how would she choose the most valuable subset of articles to be included in the issue given the page limit for the issue? This is a COP with $\mathbb{E}$ as the set of articles available, and $\mathcal{F} = \{S \subseteq 2^{\mathbb{E}} \mid \sum_{j \in S} w_j \le W\}$, that is, the collection of subsets of articles, whose total number of pages is within W. We have $c(S) = \sum_{j \in S} v_j$, $S \in \mathcal{F}$.

R. T. Rockafellar, in his classic treatise on *Convex Analysis* [151], remarks, "Convexity has been increasingly important in recent years in the study of extremum problems in many areas of applied mathematics."

This statement was made half a century ago. Since then we have seen abundant applications of convex analysis in economic theories, game theoretic modelling of economic behaviour, and convex programming (especially linear programming) formulations of industrial and business problems. However, a multitude of problems that require an understanding of non-convex extremum problems also coexist.

Any finite set with more than one element in $\mathbb{R}^d$ is a nonconvex set. Unfortunately, COP problems are optimisation problems over finite sets. So what use do we have for the tools of convex optimisation in the context of COP? In Sect. 1.4 we will see some examples and strategies to use linear optimisation over polytopes. But for now, a simple trick will do to make COP problems pose themselves as linear programming problems.

Given $\mathbb{E}$, the ground set. Let $\{0, 1\}^{|E|}$ denote the set of all $0/1$ vectors indexed by $\mathbb{E}$. Since any subset of $\mathbb{E}$ can be given by a vector, called the incidence or characteristic vector (c.v.), the collection $\mathcal{F}$ can be equivalently given by a subset F of $\{0, 1\}^{|E|}$. We could specify a combinatorial optimisation problem by giving $(\mathbb{E}, F, c)$.

Fig. 1.1 Characteristic
vector of a 5-tour

edge	(1, 2)	(1, 3)	(2, 3)	(1, 4)	(2, 4)	(3, 4)	(1, 5)	(2, 5)	(3, 5)	(4, 5)
5-tour	1	0	0	0	1	1	1	0	1	0

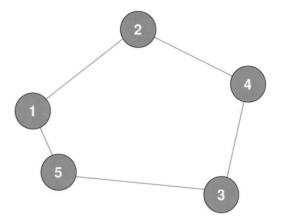

Now, let $conv(F)$ be the convex hull of elements in F. $conv(F)$ is a polytope, and a linear/convex c defined over F turns a *COP* into a linear/convex programming problem.

Example 1.3 Given the complete graph on n vertices, denoted by $K_n = (V_n = [n], E_n)$, where $E_n = \{(u, v) \in V_n \times V_n \mid u \neq v\}$ is the set of all edges. A cycle in K_n starting and ending in 1 visiting all other elements of V_n once and only once, is called a Hamiltonian cycle or n-tour. A Hamiltonian cycle can be given as a subset of edges present in the cycle.

Define the ground set $\mathbb{E} = E_n$. And F as the collection of characteristic vectors (c.v.'s) of Hamiltonian cycles in K_n. (See Fig. 1.1 for an example c.v. for a 5-city tour, given as a subset of edges.) And we are given $c : E_n \rightarrow R_+$ and $c(H) = \sum_{a \in H} c(a)$, as the cost of a Hamiltonian cycle, H.

Finding a least-cost Hamiltonian cycle is a *COP* given by $(\mathbb{E}, F, c)$. We also call this problem the symmetric travelling salesman problem (*STSP*). The polytope $conv(F)$ is called the *STSP* polytope.[2] Thus, the *STSP* problem is a linear optimisation problem over $conv(F)$.

How useful is this *COP* given by $(\mathbb{E}, F, c)$ in solving the symmetric travelling salesman problem? Not much, until we can understand the structure of this polytope well. In 1954 George B. Dantzig, Ray Fulkerson, and Selmer M. Johnson formulated their integer linear programming formulation of *STSP*. Now their formulation is known as the standard formulation [56] or *DFJ*-formulation. The variables in their formulation are 0/1 characteristic vectors representing n-tours, $x \in \{0, 1\}^{|E_n|}$. They observed that the *STSP* polytope lies on a flat as each node has degree 2 in any Hamiltonian cycle, leading to the constraint, $\sum_{e \in \delta(v)} x(e) = 2$, for all nodes, where

---

[2] *STSP* polytope for n, is denoted by Q_n in the literature.

$\delta(v)$ is the set of edges with one end as v. They realised, that this was not enough because the resulting solution might contain cycles that are not including all nodes of V_n. To eliminate them, they invented the 'subtour elimination' (SE) constraint, to be satisfied by all S, subset of $[n]$ with $2 \le |S| \le n - 2$,

$$\sum_{e \in E(S)} x(e) \le |S| - 1,$$

where $E(S)$ denotes the set of edges in K_n with both ends in the set of nodes S. However, this formulation though valid to solve the problem, has two hurdles, (1) the solutions need to be integers, and (2) the number of SE constraints is exponentially many. Therefore, in those days solving them is possible only for very small n. However, they solved a 49 city instance to optimality using many inventive ideas, as highlighted by Martin Grötschel, and George L. Nemhauser [83], such as [a] *preprocessing*, [b] *warm start*, [c] *variable fixing*, [d] *reduced cost exploitation*, [e] *cutting plane recognition*, and [f] *elements of branch-and-bound*. These ideas have found routine use later in solving many COPs. A remarkable understanding of the structure of the STSP polytope,[3] such as, the dimension [85], adjacency structure of the graph of the polytope [137] and classes of facet defining inequalities of the polytope [70], has today resulted in a branch-cut-strategy [129] used by *Concorde*, a computer code for the STSP [50] to solve not only large instances of TSP relevant to business [4], and science [148], we could even find an optimal interstellar tour visiting 109,399 stars [51].

William Cook traces the history of this famous problem and summarises the influence of this on the solution methods available for COPs in general; "The TSP has its own array of applications, but its prominence is more due to its success as an engine of discovery than it is to miles saved by travellers." He lists several known strategies, which arose from attempts to solve TSP, and are routinely used in solving difficult combinatorial optimisation problems. 'Cutting-plane method, branch-and-bound, local search, Lagrangian relaxation, and simulated annealing, to name just a few' [51].

Then what is the issue? Practically none.

But, theoretical computer science, has a simple question: is the STSP in the class of problems solvable in polynomial time? To understand this question well, we need to go back to Stephen Cook, who defined, **NP-complete** problems.

COP problems are so many and they have challenged researchers for their theoretical and practical value. As noted by Donald Knuth [106], "Many problems that once were thought to be intractable can now be polished off with ease, and many algorithms that once were known to be good have now become better." Though this summarises well the nature of COPs and their successes, we need a more precise measure of what is 'good' and what is not, when it comes to classifying these problems. For this reason, a quick peep into theoretical computer science is necessary,

---

[3] Just a glimpse of the vast literature on STSP polytope is cited here.

where decision problems are considered. For instance, we have a decision problem related to the shortest path problem.

Given n, the distance matrix D giving the distance between any pair of cities, $i, j \in [n]$ and a number l, and the question: Does there exist a path between i and j with length less than l? This is a decision problem, as we have a 'yes' answer only if there is such a path and 'no' otherwise.

Solving this decision problem is good enough to solve the corresponding optimisation problem as we can use it as a sub-problem in a binary search.

1.2 Languages, Decision Problems, Algorithms, A-Machines

'A Turing machine is a hypothetical machine thought of by the mathematician Alan Turing in 1936. Despite its simplicity, the machine can simulate ANY computer algorithm, no matter how complicated it is!'[4] With this understanding of a Turing machine, we can follow Stephen Cook's clarification of the class P problems:

Informally the class P is the class of decision problems solvable by some algorithm within a number of steps bounded by some fixed polynomial in the length of the input. Formally the elements of the class P are languages. Let Σ be a finite alphabet (that is, a finite nonempty set), and let Σ^* be the set of finite strings over Σ. Then a language over Σ is a subset L of Σ^*. Each Turing machine M has an associated input alphabet Σ. For each string w in Σ^* there is a computation associated with M with input w. ... We say that M accepts w if this computation terminates in the accepting state. Note that M fails to accept w either if this computation ends in the rejecting state, or if the computation fails to terminate. The language accepted by M, denoted $L(M)$, has associated alphabet Σ and is defined by

$$L(M) = \{w \in \Sigma^* \mid M \text{ accepts } w\}$$

We denote by $t_M(w)$ the number of steps in the computation of M on input w. ... If this computation never halts, then $t_M(w) = \infty$. For $n \in \mathbb{N}$ we denote by $T_M(n)$ the worst case run time of M; that is
$$T_M(n) = \max \{t_M(w) \mid w \in \Sigma^n\}$$
where Σ^n is the set of all strings over Σ of length n. We say that M runs in polynomial time if there exists k such that for all n, $T_M(n) \leq n^k + k$. Now we define the class P of languages by P = $\{L \mid L = L(M)$ for some Turing machine M which runs in polynomial time$\}$ [48].

To understand what follows in this book, it will be enough to remember class P as the class of problems solvable by a digital computer in time polynomial in the input length of a problem instance given as a $0-1$ string.

Another concept we need to understand is how one decision problem is transformable into another decision problem.

---

[4] https://www.cl.cam.ac.uk/projects/raspberrypi/tutorials/turing-machine/one.html.

Definition 1.2 Given two decision problems $P_1 \subseteq \{0, 1\}^*$ and $P_2 \subseteq \{0, 1\}^*$, we say P_1 is transformable to P_2 if there is a function f such that

1. $f : \{0, 1\}^* \to \{0, 1\}^*$,
2. there is an algorithm to compute f in polynomial-time, and
3. $f(x) \in P_2$ if and only if $x \in P_1$.

In the class NP of decision problems (languages)[5] we have a polynomial time verification scheme or certificate to check a given instance is a 'yes' instance (that is, the given string belongs to the language, L).

For example, in the decision problem of deciding whether a graph is Hamiltonian, producing a Hamiltonian Cycle (that is, an encoding of H) is a certificate for the given graph instance is a 'yes' instance. Because we can verify easily whether the nodes in H cover all nodes in G, and whether the edges in H produce a cycle in G connecting all nodes, by just tracing the edges starting from any node of G and tracing the edges connected to that node in H. Thus Hamiltonian Cycle problem is in the class NP of decision problems.

Seminal work by Stephen Cook [49] is defining the NP-complete class which is a subclass of NP defined as follows:

Definition 1.3 A language L is NP-complete if and only if L is in NP, and L' is transformable to L for every language L' in NP.   ♣

This implies, to include a new language L in NP-complete class, we first show that [1] we have a polynomial verification scheme for L and then show that [2] every language $L' \in$ NP is transformable to L. An important consequence of defining NP − completeness is the result:

Theorem 1.1 (Cook's Theorem)

1. *If L_1 is transformable to L_2 and $L_2 \in$ P then $L_1 \in$ P.*
2. *If L_1 is NP-complete, $L_2 \in$ NP, and L_1 is transformable to L_2 then L_2 is NP-complete.*
3. *If L is NP-complete and $L \in$ P, then P $=$ NP.*

Immediately after Cook's result [49] appeared, Richard Karp [101] showed many problems of importance, to graph theorists and operations research community are in NP-complete class, including the integer programming problem. Since then

---

[5] Here, NP stands for non-deterministic polynomial. An earlier characterisation of this class of problems uses what are called non-deterministic Turing machines.

NP-complete subclass of NP is known to consist of many decision problems related to difficult combinatorial problems. (See Michael Garey and David Johnson's celebrated collection of NP-complete problems [76].)

1.3 A New Beginning

Combinatorial optimisation problems are of our interest. Their complexity can be related to that of the corresponding decision problems studied. If a problem is not in NP, like in the case of an optimisation problem, but it is as hard as some NP-complete problem, then it is called NP-hard. A typical problem in this class is the symmetric travelling salesman problem.

1.3.1 How Hard Is the Symmetric Travelling Salesman Problem?

Given an instance of *STSP* to answer the decision version of the problem, we can use the *Method-of-Exhaustion*:

1. Set $p^* = \emptyset$, *least-cost* = l.
2. Select a member p in $\mathcal{F} \neq \emptyset$.
3. If $c(p) \leq$ *least-cost*:
       set *least-cost* = $c(p)$, $p^* = p$.
       Answer: 'yes', Output: p^*, $c(p^*)$, Stop.
   Otherwise :
       delete p from $\mathcal{F}$.
4.   If $\mathcal{F} = \emptyset$:
          Answer: 'no', Stop.
       Otherwise:
          Repeat Step 2.

Yes, this method is the infamous Brute force algorithm [47], also called the trial and error method. However, I prefer to use the name *Method-of-Exhaustion* for what it does and for some *COP* problems we have only this approach for solving; or some of its variants are proposed.

The variants of the *Method-of-Exhaustion* include divide-and-conquer strategies, Richard Bellman's dynamic programming [28] approach, branch-bound-schemes or some way to exhaust the arbitrarily many choices present at different stages of the search process, by cutting off or pruning some subset of $\mathcal{F}$. They do reduce the computational burden, but the complexity class does not change, as we shall see. However, while using any *Method-of-Exhaustion* or it's variant to solve a *COP* problem, we may have found the best object of interest very quickly in the search, but we have no clue that it is the best. So the rest of the computational burden is in proving the optimality of the solution at hand. I can call this *cost of establishing optimality*. This cost is reducible depending on our polyhedral knowledge of the collection of objects of interest and the behaviour of the cost function.

For instance, it is interesting to note Richard Bellman applied the dynamic programming approach for solving *STSP* [28] around the same time Michael Held and Richard Karp [91] gave their version of the dynamic programming formulation.

> **Definition 1.4** For every non-empty subset $S \subseteq \{2, \dots, n\}$ and for every city $i \in S$, let OPT[$S; i$] denote the length of the shortest path that starts in city 1, then visits all cities in $S - \{i\}$ in arbitrary order, and finally stops in city i. So, OPT[$\{i\}; i$] = $d(1, i)$. ♣

Then we can arrive at the recursive equation

$$\text{OPT}[S; i] = \min\{\text{OPT}[S - \{i\}; j] + d(j, i) : j \in S - \{i\}\}.$$

The optimal travel length is given as the minimum value of OPT [$\{2, \dots, n\}; j$] + $d(j, 1)$ over all j with $2 \leq j \leq n$.

Noticing the value OPT [$S; i$] can be computed in time proportional to $|S|$, we have a time complexity of the order of $n^2 \times 2^n$ instead of $n!$.

Gerhard Woeginger [164] while surveying exact algorithms for NP-hard problems based on dynamic programming approaches, comments on Held & Karp's dynamic programming formulation of *STSP* as follows:

> This result was published in 1962, and from nowadays point of view almost looks trivial. Still, it yields the best time complexity that is known today.

 Note *(Observation 1) Given a COP, lesser we employ the properties of [i] the collection of objects of interest, $\mathcal{F}$ or [ii] the function c, higher the likelihood that some form of* Method-of-Exhaustion *is employed in solving the problem.*

But can we do better?

This book is on a combinatorial object I call *pedigree*. Before I discuss why I chose to study these objects and their relevance to the foregoing discussions on the complexity of *COP* problems, I briefly introduce the pedigrees.

1.3.2   What Are Pedigrees?

Let the set of first n natural numbers, $\{1, \dots, n\}$ be denoted by $[n]$. Any ordered 3-element set, $\{i, j, k\}$, $1 \leq i < j < k \leq n$, for $n > 3$, is called a triangle (triplet, 3-tuple or 2-simplex are also used to refer to a triangle). The complete collection of triangles is denoted by Δ. We partition Δ based on the maximum element in any $u \in \Delta$. Thus, $u = \{1, 2, 3\}$ is the unique element in its class Δ^3. Similarly, we understand $\Delta^k = \{ \{i, j, k\} \mid 1 \le i < j < k \}\}$, for $k = 4, \ldots, n$, with $|\Delta^k| = (k-1)(k-2)/2$.

Given $u \in \Delta^l$, and $v \in \Delta^k$, $l < k$, for $k = 4, \ldots, n$, we say u is a generator of v or equivalently we say, v is a descendant of u if

1. $u \cap v = \{v \setminus \{k\}\}$, and
2. $u \cap v \ne \{u \setminus \{l\}\}$, except when $l = 3$. $u \cap v$ is called the common facet or edge connecting u and v.

Note $\{1, 2, 3\}$ *has no generator. All* $v = \{i, j, k\}$ *with* $\{i, j\} \in \{1, 2, 3\}$ *have* $\{1, 2, 3\}$ *as generator. But* $u = \{i, j, l\}$ *cannot be a generator of* $v = \{i, j, k\}$, *for* $k > l > 3$. *For* $v = \{i, j, k\}$ *with* $\{i, j\}$ *not in* $\{1, 2, 3\}$ *we have the set of generators for* v *given by* $\{\{r, i, j\}, 1 \le r \le i - 1\} \cup \{\{i, s, j\}, i + 1 \le s \le j - 1\}\}$.

For instance, $u = \{1, 2, 3\} \in \Delta^3$, is a generator of $v = \{2, 3, 7\} \in \Delta^7$ and $\{2, 3\}$ is the common edge. On the other hand, $v = \{2, 3, 7\}$ is not a descendant of $u = \{2, 3, 5\}$ though $\{2, 3\}$ is the common edge, it violates stipulation [2]. Any subset P of Δ is called a *spanning set or represent* if $P \cap \Delta^k \ne \emptyset$, for $k = 3, \ldots n$. For instance, given $n = 5$, $P = \{\{1, 2, 3\}, \{1, 2, 4\}, \{2, 3, 4\}, \{2, 3, 5\}, \{1, 4, 5\}\}$ is a spanning set. However, a subset of P, $P^* = \{\{1, 2, 3\}, \{2, 3, 4\}, \{1, 4, 5\}\}$ is also a spanning set. Thus we may look for minimal represents. Also, the elements in P other than $\{1, 2, 3\}$ have a generator in P. However, we can not say this about P^*, as $\{1, 4, 5\}$ has no generator in P^*.

Now consider the directed graph $D = (V, A)$ where $V = \Delta$ and $A = \{(u, v) | u, v \in V \text{ and } u \text{ is a generator of } v\}$.

Definition 1.5 (*Pedigree*) A spanning set P in D is called a *Pedigree*, if and only if it is a tree,

1. rooted at $\{1, 2, 3\}$,
2. has exactly one element from each Δ^k, $k = 3, \ldots, n$, so $|P| = n - 2$, and
3. every element of P other than the root has a generator in P, and the common edges are all distinct. ♣

Definition 1.5 is clarified using Example 1.4.

Example 1.4 [a] Given $n = 6$, $P = \{\{1, 2, 3\}, \{2, 3, 4\}, \{1, 3, 5\}, \{1, 2, 6\}\}$ is a pedigree (see Fig. 1.2).

[b] Given $n = 7$, $P = \{\{1, 2, 3\}, \{2, 3, 4\}, \{2, 4, 5\}, \{4, 5, 6\}, \{4, 6, 7\}\}$ is a pedigree (see Fig. 1.3).

[c] Given $n = 7$, $P = \{\{1, 2, 3\}, \{2, 3, 4\}, \{2, 4, 5\}, \{4, 5, 6\}, \{2, 4, 7\}\}$ is not a pedigree. As $\{2, 4, 5\}, \{2, 4, 7\}$ have $\{2, 3, 4\}$ as a generator, and the common edges are not distinct.

Fig. 1.2 Pedigree:
$P = \{\{1, 2, 3\},$
$\{2, 3, 4\}, \{1, 3, 5\},$
$\{1, 2, 6\}\}$

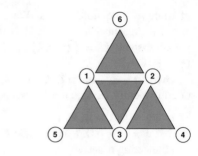

Fig. 1.3 Pedigree:
$P = \{\{1, 2, 3\},$
$\{2, 3, 4\}, \{2, 4, 5\}, \{4, 5, 6\},$
$\{4, 6, 7\}\}$

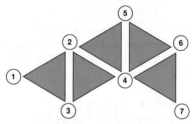

Suppose we are given a nonnegative rational function, $c : \Delta \rightarrow Q_+$. We can define the cost of any subset P of Δ as $cost(P) = \sum_{u \in P} c(u)$. In this book, I shall outline how to efficiently solve Problem 1.1.

Problem 1.1 (*Pedigree Optimisation*)
Given: $n > 3$, consider the directed graph $D = (\Delta, A)$ where A is the collection of all $(u, v) \in \Delta \times \Delta$ such that u is a generator of v, and the cost function, $c : \Delta \rightarrow Q_+$.
Find: P^* a pedigree that minimises $c(P)$ over all pedigrees in D.

Before I discuss this problem in detail and provide a solution approach, I will quickly go through some relevant milestones in combinatorial optimisation, the use of linear programming concepts in such problems, not as a survey or review or tutorial, but to capture some *clues* from the stalwarts who made remarkable shifts in the field, for designing the strategies and picking up tools, for solving the pedigree optimisation problem efficiently.

1.4   Insightful Strategies and Inexpensive Slingshots

This section browses through a bunch of interrelated problems and solution approaches stemming from the transportation problem studied by Koopmans and Reiter [107] ending with the class of problems called combinatorial linear programming problems studied by Éva Tardos [159]. At the end of this section, I would have gathered sufficient clues for designing a solution approach for the pedigree optimisation problem.

Imagine two ports between which cargo is transported by ships, either carrying cargo or sailing empty from one port to the other. It makes economic sense to minimise the number of empty trips made by the ships. Koopmans and Reiter study such a problem in [107]. They remark in their article, 'A model on Transportation', (in Sect. 2.11):

> There is an interesting analogy, with differences, between the problem of minimising the amount of shipping in use for a given transportation program and the distribution of (direct) current in a network of electrical conductors to which given electromotive forces are applied at specified points. The latter problem, treated by Kirchhoff [1847], provided the stimulus for the mathematical investigation of linear graphs.

The analogy is brought out by the following list of reinterpretations of the symbols used above (in [107]).

Transportation Model	Interpretation in Symbol	Electrical Network
Ports	$i = 1, \ldots, n$	Connection points of conductors
Routes	(i, j)	Conductors
Empty sailing time	s_{ij}	Resistance
Flow of empty ships	x_{ij}	Electrical current
Net shipping surplus	b_i	Net current made to flow into the network from outside
Locational potential	p_i	Negative of electrical potential

What is even more interesting is a footnote found in [107], "The cultural lag of economic thought in the application of mathematical methods is strikingly illustrated by the fact that linear graphs are making their entrance into transportation theory just about a century after they were first studied in relation to electrical networks, although organised transportation systems are much older than the study of electricity."

 Note *(Observation 2) Firstly, this is an example of reducing a* new *problem involving economic activity to a known problem in physics using an analogy. However, we might employ other means for this visualisation, as long as it is easy to show, one problem is an instance of another problem.*

Secondly, a problem seeking whole number solutions (such as the number of empty trips) can use models that can find continuous solutions (like current flow in an electrical network).

Thirdly, this article brings out the importance of finding solutions that are whole numbers in everyday economic activities and solving decision problems. (Like, make or buy situations faced by manufacturers, or deciding on the number of members in a committee.)

Ray Fulkerson [75] notes, "The transportation problem was first formulated by P. L. Hitchcock in 1941. He also gave a computational procedure, much akin to the general simplex method, for solving the problem." Hitchcock's formulation is as given in the next page:

Problem 1.2 (*Hitchcock Problem*) Find an $m \times n$ array of numbers $x = (x_{ij})$, $i = 1, 2, \ldots, m$ and $j = 1, 2, \ldots, n$, that minimises $\sum_{i,j} c_{ij} x_{ij}$ subject to the constraints

$$\sum_j x_{ij} = a_i, \text{ for all } i \in [m],$$

$$\sum_i x_{ij} = b_j, \text{ for all } j \in [n],$$

$$x_{ij} \geqq 0$$

where a_i, b_j, c_{ij} are given nonnegative integers.

This is an instance of a linear programming problem having the form,

$$\min \left\{ c^\top x : Ax = b, x \in \mathbb{R}_+^n \right\},$$

where A is a $m \times n$ real matrix, $b \in \mathbb{R}^m$, and $c \in \mathbb{R}^n$.

While tracing the origins of his brainchild, the simplex method, George B. Dantzig [59], remarks on his very innovative move to use linear inequalities for modelling economic activities.

Curiously enough up to 1947 when I first proposed that a model based on linear inequalities be used for planning activities of large-scale enterprises, linear inequality theory had produced only forty or so papers in contrast to linear equation theory and the related subjects of linear algebra and approximation which had produced a vast literature. Perhaps this disproportionate interest in linear equation theory was motivated .... by the belief that linear inequality systems would not be practical to solve unless they had three or less variables.

George B. Dantzig [57] noticing the special structure of the matrix A, in Hitchcock's formulation of the transportation problem (that is, [i] it has $m + n$ rows and $m \times n$ columns, and [ii] each column has exactly two 1's.) modified the simplex method to take advantage of the same. If the right-hand side in the Hitchcock problem is all $1's$, and $m = n$, we have the bipartite matching problem, also called, the assignment problem.

On solving this problem, using the linear programming model and the simplex method, Dantzig makes the following comment [58]:

The tremendous power of the simplex method is difficult to realise. To solve by brute force the Assignment Problem which I mentioned earlier would require a solar system full of nano-second electronic computers running from the time of the big bang until the time the universe grows cold to scan all the permutations in order to be certain to find the one which is best.

An entirely different approach to solving the Hitchcock problem stems from an observation made by Ray Fulkerson. While commenting on Hitchcock's problem Fulkerson [75] remarks, "From a mathematical point of view, perhaps the most interesting distinguishing feature of the transportation problem is that it provides an approach to some problems which at first appear to be purely combinatorial."

Soon after that, as a further extension of the analogy of electrical networks to the transportation problem, Ford and Fulkerson [72] approached the Hitchcock problem as a particular instance of the flow problem in networks, in general. They remark:

> The problem arises naturally in the study of transportation networks; it may be stated in the following way. One is given a network of directed arcs and nodes with two distinguished nodes, called *source* and *sink*, respectively. All other nodes are called *intermediate*. Each directed arc in the network has associated with it a nonnegative integer, its flow capacity. Source arcs may be assumed to be directed away from the source, sink arcs into the sink. Subject to the conditions that the flow in an arc is in the direction of the arc and does not exceed its capacity, and that the total flow into any intermediate node is equal to the flow out of it, it is desired to find a maximal flow from source to sink in the network, i.e., a flow which maximises the sum of the flows in source (or sink) arcs.

More formally, given a directed graph $G = (V, A)$ with arc weights $c \in \mathbf{Z}_+^{|A|}$, source $s \in V$, sink $t \in V$, Ford–Fulkerson's problem can be stated as

Problem 1.3 (*Max-flow Problem*) Let an st-flow (flow) f be a function that satisfies:
$$0 \leq f(e) \leq c(e), \text{ for every } e \in A \text{ [nonnegativity, and capacity restrictions]}$$
$$\sum_{e \text{ into } v} f(e) = \sum_{e \text{ out of } v} f(e), \quad \text{for intermediate } v \in V \text{ [flow conservation]}$$
Let the value of a flow f be defined as $\text{val}(f) = \sum_{e \text{ out of } s} f(e) - \sum_{e \text{ into } s} f(e)$
Find: A flow f with maximum value.

A naive approach for solving Max-flow problems is to start with some st-flow and then look for a st-path P where each edge has $f(e) < c(e)$, if one exists increase flow along path P, by $\delta = \min_{e \in P}(c(e) - f(e))$. Repeat this until you cannot find any st-path. But we have no evidence that this will be the maximum flow. The method proposed by Ford and Fulkerson to solve the Max-flow problem goes by the name Ford–Fulkerson labelling method. This method is based on two inventive ideas: [1] residual network and [2] flow augmenting path.

Definition 1.6 (*Residual Graph*) With respect to a feasible flow f for a *Max-flow* problem, we define a mixed graph $G_f = (V, A_f)$ where V is as given in the *Max-flow* problem, and

$$A_f = A_{forward} \cup A_{reverse}, \text{ where}$$

$$A_{forward} = \{(u, v) \mid f(a) < c(a), a = (u, v) \in A\}, \text{ and}$$
$$A_{reverse} = \{(v, u) \mid f(a) > 0, a = (u, v) \in A\}.$$

We define the capacity $c_f(a)$ for arcs in A_f as

$$c_f(a) = \begin{cases} c(a) - f(a) & \text{if } a \in A_{forward} \\ f(a) & \text{if } a \in A_{reverse}. \end{cases}$$

♣

Definition 1.7 (*Augmenting Path*) An augmenting path is a simple st-path in the residual network G_f. Let the minimum capacity along the augmenting path be δ. Now we can increase the flow along the forward arcs by δ and decrease the flow along the reverse arcs by δ obtaining a new improved flow f' having value $\text{val}(f) + \delta$.   ♣

From the definition of the residual graph G_f, it is clear that if there is a st-path in the residual graph we can have a positive flow along this path from s to t. And the value of that flow can be equal to the minimum capacity of any arc in that path. And the suggested flow change does not violate any of the restrictions on an st-flow in G. Therefore, we have an improved solution. Here we can state the algorithm given in [73] as follows:

Algorithm 1 Max-Flow Algorithm (Ford and Fulkerson [73])

Input: Directed graph $G = (V, A)$ with arc weights $c \in \mathbf{Z}_+^{|A|}$, source $s \in V$, sink $t \in V$.
Output: Maximum (s, t)-flow f.
STEP 1: Start with any st-flow f. ($f = 0$ will do.)
STEP 2: Find an augmenting path P in the residual graph, G_f.
if such a path P exists:
    Find the augmented flow f' given by $f'(a) = f(a) + \delta$ if $a \in A_{forward}$,
    otherwise $f'(a) = f(a) - \delta$. Repeat STEP 2.
else STOP.

Notice that the choice of the augmenting path, if one exists, is arbitrary (non-deterministic) in the Ford–Fulkerson algorithm. Edmonds and Karp [68] give an example (see Fig. 1.4 and the example below.) where the algorithm may take an arbitrarily large number of steps, and prove Theorem 1.2 which gives a modification of the Ford–Fulkerson algorithm.

In this example, we have a network with 4 nodes including *source* and *sink*, all arcs except (u, v) have capacity M a large integer, and (u, v) has capacity 1. According to the Ford–Fulkerson algorithm, if we choose the st-path $s \rightarrow u \rightarrow v \rightarrow t$ as our augmenting path every time, the number of iterations will be $2M$.

Fig. 1.4 Edmonds–Karp's example. Capacities are shown along the arcs

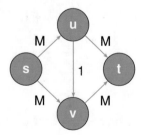

Note *(Observation 3) Non-determinism in choice can lead to a heavy computational burden.*

> **Theorem 1.2** *If in the labelling method for finding a maximum flow in a network on n nodes, each flow augmentation is done along an augmenting path having the fewest arcs, then a maximum flow will be obtained after no more than $\frac{1}{4}\left(n^3 - n\right)$ augmentations.*

There are many improved algorithms available for finding maximum flows in networks [108], but for our use Edmonds–Karp's modification of the labelling method is sufficient.

When the set of intermediate nodes can be partitioned into two parts (origins and destinations) such that there are only arcs between origins and destinations, we have a sub-class of network flow problems, called bipartite flow problems. We can easily see that the Hitchcock problem is a bipartite flow problem, with a source 's' connected to each of the origin nodes by an arc with the availability as capacity and a sink 't' connected to each of the destination nodes with demand as capacity respectively, as shown in Fig. 1.5.

Bipartite flow problems are well-solved problems. Next, I shall elaborate on what I mean by well-solved. For instance, Claude Berge considers three graph problems in [30] and the third problem is

Problem 1.4 *(Berge's Third Problem)* Given a finite graph $G = (V, E)$, a set of edges $M \subset E$ is said to be a matching if two edges of M have no vertex in common. Construct a matching with the maximum number of elements.

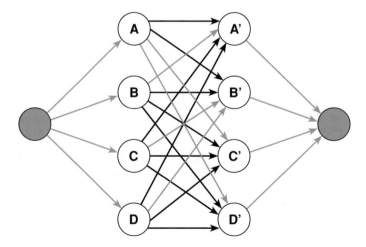

Fig. 1.5 Bipartite flow problem

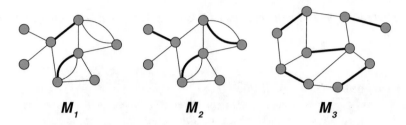

Fig. 1.6 Examples of matchings. M_1 and M_2 are in the same graph. Both are maximal. M_3 is a perfect matching for the graph

A matching is *maximal* if we add any more edge to it, it will not be a matching. A matching is maximum if the number of edges in the matching is maximum. Figure 1.6 gives examples of matchings in graphs. Berge gives a necessary and sufficient condition for recognising whether a matching is maximum and provides an algorithm for Problem 1.4.

Consider a matching M; if any edge $u \in M$ is called *strong*, edges $E \setminus M$ are called *weak*. An *alternating chain* is a chain that does not use the same edge twice and is such that for any two adjacent edges in the chain one is strong and the other is weak. A vertex x which is not adjacent to a strong edge is said to be *neutral*. In Fig. 1.6, in M_1 we can find an alternating chain, but not in M_2.

> **Theorem 1.3** (Berge's Theorem 1 [30]) *A matching M is maximum if and only if there does not exist an alternating chain connecting a neutral point to another neutral point.*   ♡

Berge states,

Theorem 1 suggests the following procedure for solving Problem 3[6]:

1. Construct a maximal matching M,
2. Determine whether there exists an alternating chain P connecting two neutral points. (The procedure is known.)
3. If such a chain exists, change M into $(M \setminus P) \cup (P \setminus M)$, and look again for a new alternating chain; repeat Step 2, Otherwise, that is, such a chain does not exist, and M is maximum. Stop.

Berge's suggestion for solving his Problem 3 has some resemblance to *Method-of-Exhaustion*, as there is some arbitrariness in finding the next matching to consider, although the theory he has built has discovered the structure of the collection of matchings well. Thus Richard Karp notes, "To implement this approach, it is neces-

---

[6] Problem 1.4 restated in this chapter.

sary to have an efficient method of finding an augmenting path (alternating chain) or determining that none exists" [102].

Note *(Observation 4) Discovering the properties and structure of the collection of objects of interest, though necessary, is not sufficient to escape the curse of enumeration that requires some kind of* Method-of-Exhaustion.

A great milestone in computational complexity as well as in combinatorial optimisation problems is the work of Jack Edmonds, which he has organically entitled, 'PATHS, TREES, AND FLOWERS'. Edmonds builds on the existing theory to invent a method that will avoid the non-determinism present in Berge's approach. From this work, we can make three important observations: [1] identifying traps of non-determinism in an approach, [2] inventing preventive strategies based on observed pitfalls/obstacles, [3] solving an integer optimisation problem using a linear programming formulation, simplex method and LP duality.

In [64] Edmonds notes:

> In fact, he (Berge) proposed to trace out an alternating path from an exposed vertex[7] until it must stop and then, if it is not augmenting, to back up a little and try again, thereby exhausting possibilities. His idea is an important improvement over the completely naive algorithm. However, depending on what further directions are given, the task can still be one of exponential order, requiring an equally large memory to know when it is done.

After defining what are [1] a planted tree, [2] an augmenting tree, [3] a flowered tree and [4] a Hungarian tree, Edmonds is able to show, "For a matching M in a graph G, an exposed vertex is a planted tree. Any planted tree J(M) in G can be extended either to an augmenting tree, or to a flowered tree, or to a Hungarian tree (merely by looking at most once at each of the edges in G which join vertices of the final tree)." His Theorem 4.8 makes an important statement on searching, "The algorithm which is being constructed is efficient because it does not require tracing many various combinations of the same edges in order to find an augmenting path or to determine that there are none."

In Digression Edmonds remarks, 'An explanation is due on the use of the words "efficient algorithm." First, what I present is a conceptual description of an algorithm and not a particular formalised algorithm or "code."'

Edmonds' epoch-making algorithm and the paper have many clear expositions on the cited books, notes, and articles on the Blossom algorithm, like [67, 108, 154, 157]. As pointed out by Pullyblank [144], we have seen the birth of the field now known as *Polyhedral Combinatorics*. Polyhedral combinatorics is in the confluence of combinatorial optimisation, linear programming, and computational efficiency.

Our purpose of going through Edmonds' classic is to observe for clues for inventing 'non-exhausting methods' for difficult *COPs*.

---

[7] What Berge calls a 'neutral node' is called here an exposed vertex by Edmonds.

1.4.1 Tools from Past, Slingshots for Attack

In this book, I will consider a transportation problem called *Forbidden arc trans-portation problem*, in which some origins and destinations are not connected by arcs, they are called forbidden arcs. In Fig. 1.5 these are shown as 'black' arcs. They are bipartite network flow problems, with some capacities of arcs being zeros, and others having no or some capacity restrictions. In addition, I will also be discussing flow problems involving multiple commodities, where each commodity flow can be rational, instead of an integer, for each arc in the network. For each commodity k we have source, sink pairs s_k, t_k, capacity restrictions $f_k(a) \leq c_k(a)$ for each commodity, apart from the capacity restriction on the total flow, $f(a) = \sum_k f_k(a) \leq c(a)$ on arc a in the network. Solving the multicommodity flow problem is by itself an interesting pursuit as the problem becomes difficult, soon we have two commodities and we seek integer solutions. Luckily, in our application, we seek only rational solutions and not integer solutions to the multicommodity flow problems that we solve. These flow problems are well-solved optimisation problems.

In Sect. 1.2 we defined the class P, as problems that are solvable in polynomial time in the input length of an instance. I did not say much about what this input length means. For a linear programming problem, we have the matrix A, the right-hand side, b, and the objective coefficients, c to be encoded as a 0, 1 string as input. So the length of this string is what we call the input size/length.

The polynomial solvability of LP remained an unanswered question until February 1979. The inventive idea of L. G. Khachiyan [103] to cleverly use Shor's [158] ellipsoid method for solving a linear inequality system ($Ax \leq b$) changed that. Especially, Khachiyan made it explicit that working with finite precision, his method is polynomial in the input size of an LP instance. Although the Ellipsoid algorithm is useful in proving the polynomial solvability of the LP problem, its performance is far behind that of the champion—the simplex method, which is not of polynomial complexity. (See Chap. 2 for an example and other details on LP and simplex method, or the books cited therein.) So researchers have to refine polynomial solvability further and we have what is called the notion of strongly polynomial algorithms. Martin Grötschel, László Lovász, and Alexander Schrijver in their celebrated work [82] made the best use of the ellipsoidal method in showing many graph problems are strongly polynomial. For defining what is a strongly polynomial algorithm, we need to understand what is the dimension of the input means.

Definition 1.8 (*Dimension of the input*) We call the dimension of the input as the number of data items in the input (that is, each item is considered to add one to the dimension of the input). The size of a rational number p/q is the length of its binary description (i.e., $size(p/q) = \lceil log_2(p + 1) \rceil + \lceil log_2(q + 1) \rceil$ where $\lceil x \rceil$ denotes the smallest integer not less than x). The size of a rational vector or matrix is the sum of the sizes of its entries. ♣

Definition 1.9 (*Strongly Polynomial*) An algorithm is strongly polynomial [82] if

A. it consists of the (elementary) arithmetic operations: addition, comparison, multiplication and division;
B. the number of such steps is polynomially bounded in the dimension of the input (as defined above);
C. when the algorithm is applied to rational input, then the size of the numbers occurring during the algorithm is polynomially bounded in the dimension of the input and the size of the input numbers. ♣

Next, we shall discuss what are combinatorial linear programming problems.

Definition 1.10 (*Combinatorial LP*) A class of linear programming problems is called *combinatorial* if the size of the entries of the matrix A is polynomially bounded in the dimension of the problem [159]. ♣

Éva Tardos [159] was the first to provide a strongly polynomial algorithm that solves a combinatorial LP. That is, the number of arithmetic steps used by Tardos's algorithm depends only on the dimension of the matrix A, but is independent of both b and c. A quick look at the problems we have so far encountered in this section, like the transportation problem, maximal flow problem, and the multicommodity flow problem are all having a special kind of matrix A. They are sparse, that is, many elements are 0, and other elements are $+1$ or -1. Therefore, it is easy to observe that these problems are combinatorial linear programming problems. Therefore, we are assured of a strongly polynomial algorithm for solving these problems, using Tardos's theorem.

This theoretical result is useful in establishing results on the complexity of the approach outlined in the book for the main problem, namely, the pedigree optimisation problem.

1.5 Strategies for Avoiding Non-determinism

From the observations made looking at milestone discoveries and inventive algorithms proposed by George B. Dantzig, T. C. Koopmans, L. R. Ford Jr., D. R. Fulkerson, Richard Bellman, Ralph Gomory, Claude Berge, Jack Edmonds, Richard. M. Karp, Steven Cook and Éva Tardos, I have liberally *rephrased* certain do's and don'ts for us to follow while setting out to search for a robust strategy for finding a polynomial algorithm for a difficult combinatorial optimisation problem.

Do List:

1. Leave the flat world (created by linear equations) and enter a new world of half spaces (to formulate your models). [from G. B. Dantzig]
2. Change the problem into a multistage one and reap the benefits of the principle of optimality. [from R. Bellman]
3. You will be shocked by the benefits of Flow analogies. [from T. C. Koopmans]
4. Cuts are too many to search, go with the flow, getting minimally cut. [from L. R. Ford Jr. and D. R. Fulkerson]
5. Hitchcock problems are everywhere. Look for them even if they are forbidden. [from D. R. Fulkerson]
6. There is no match for understanding the structure if you want to solve a problem. [from C. Berge]
7. Be determined to eliminate non-determinism, the algorithm will blossom automatically. [from J. Edmonds]
8. Matrix matters more than resources or objectives, for the strength of an algorithm. [from: Éva Tardos]

Don't List:

9. Don't let non-determinism (arbitrary choice) creep in anywhere in an algorithm. [from Stalwarts in one voice]

In this book, I have applied these clues in designing an efficient algorithmic framework for solving the pedigree optimisation problem, especially in discovering the structure of the pedigree polytope and in inventing an algorithm *following the Don't clue.* For instance, [1] though I have not left the 1-simplex (graph) world completely I am working with the 2-simplex or triangular world; [2] I have considered a multistage decision formulation of the symmetric travelling salesman problem; [3] I have discovered the structure of pedigree polytope from many different perspectives; [4] all the problems solved are combinatorial LP problems, some are just bipartite flow problems; [5] the framework I suggest for checking the membership in pedigree polytope is such that non-determinism is completely avoided; [6] the validation of the framework and the polynomial upper bound, use the properties of the pedigree polytope and the inventive use of flow problems solved; and most importantly [7] an analogy to protein folding helps in viewing a vector from $\mathbb{Q}^{|\Delta|}$ folded on to a two-dimensional lattice, and then recursively constructing and solving flow problems in layered networks.

The structure of the remaining chapters of the book is outlined next.

1.6   Structure of the Book

Chapter 2 contains preliminaries on concepts from polytopes, convex sets, graph theory, linear programming, flows in networks, and other areas of relevance. However, one can skip Chap. 2 except for the basic notations and Subsection on Forbidden Arc Transportation problems.

Chapter 3 introduces the motivations for studying pedigrees and their connection to tours. Some of you might have already noticed that in Figs. 1.2 and 1.3 the boundaries of these pedigrees are Hamiltonian cycles. Dynamic programming-inspired multistage insertion (*MI*) formulation for the symmetric travelling salesman problem is given and its properties are explored. It is established that the subtour elimination polytope of Dantzig, Fulkerson and Johnson is equivalent to the *MI* relaxation polytope.

Chapter 4 introduces the pedigree polytope and characterises it. Furthermore, the adjacency structure of the pedigree polytope is explored by obtaining a strongly polynomial time algorithm to test the non-adjacency of pedigrees.

Chapter 5 deals with the membership problem of pedigree polytopes. An algorithmic framework is given for checking the membership in the pedigree polytope.

Chapter 6 demonstrates that this framework is implementable in strongly polynomial time. An example illustrates the steps in the framework.

Chapter 7 brings out the crucial implication of the results found in Chaps. 5 and 6. After checking for the necessary technical details for using the polynomial membership checking algorithm obtained in Chap. 5 to construct a violated facet of the polytope suggested by Maurras [121], we can see the pedigree optimisation problem is polynomially solvable using results from [82]. Thus this chapter using the connection between pedigree optimisation and *STSP* problem shows a problem in the NP-complete class is in P. And as an immediate consequence of Cook's Theorem (Theorem 1.1) we have NP = P, a result not expected by many researchers.

Chapter 8 is the concluding chapter of the book; it identifies some immediate future research areas. This chapter reports some encouraging results from early computational experiments done (much prior to the results in Chaps. 5–7 were obtained). These experiments were carried out using small to medium-sized instances from the TSPLIB [148] list of problems. And so, solving larger instances of *MI*-relaxation problem is needed. This chapter outlines the hypergraph flow formulation and the Lagrangian relaxation approach to solving larger instances of the *MI*-relaxation problem.

Overall the chapters flow in the given order, namely, the prelude, the basics, the motivation, the polytope, the membership problem, its complexity, the consequences and the future directions. Since the book introduces a new 0/1 polytope having interesting properties, the study of Pedigree Polytopes by itself might be a sufficient reason for those researchers involved in studying polytopes. The connection between pedigrees and Hamiltonian cycles studied in the book might be of interest to operations

researchers working in polyhedral combinatorics of the symmetric travelling salesman problem. The efficient checking of the membership in the pedigree polytope,
with its consequences, will attract theoretical computer science researchers, working
on computational complexity, irrespective of their held beliefs concerning $\mathsf{NP} = \mathsf{P}$.

On the other hand practitioners and applied researchers would like to explore new
algorithmic possibilities opened by the *MI*-formulation and the related research. So
in Sect. 1.6.1 we deal with the different needs and interests of the readers and suggest
different ways of using the book.

1.6.1   Pathways to Read the Book

*Pedigree Polytopes—New insights on Computational Complexity of Combinatorial
Optimisation Problems* can be read differently depending on the intended purpose
of the reader. The different purposes and pathways to read the book are (shown in
Fig. 1.7) as follows:

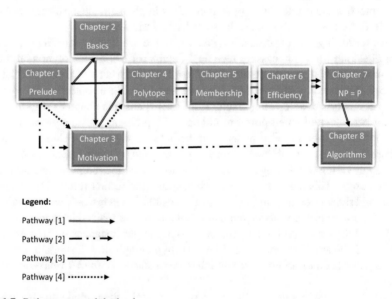

Fig. 1.7 Pathways to read the book

1 to know what is in the book -
   Pathway 1: This chapter through Chap. 8.
2 to understand new algorithmic possibilities -
   Pathway 2: This chapter, Chaps. 3 and 8.
3 to check the proof of $NP = P$ -
   Pathway 3: This chapter, Chaps. 4 through 7.
4 to understand the pedigree polytope and its properties -
   Pathway 4: This chapter, Chaps. 3 through 6.

Chapter 2 can be skipped or used as needed, but definitions, concepts, and methods from Sects. 2.5 through 2.7 are used in later chapters.

Chapter 2
Notations, Definitions and Briefs

Algorithmic solutions to combinatorial optimisation problems owe a lot to the successful use of approaches that assert whether a given set of linear inequalities is empty or not. Linear programming formulations, concepts like duality, and complementary slackness, and search approaches like the simplex method, its variations and interior point methods are employed in designing efficient algorithms for combinatorial optimisation problems. An important work that needs special mention in this area of polyhedral combinatorics is the book, *Geometric algorithms and combinatorial optimisation*[1] by Grötschel, Lovász, and Schrijver [82]. Plenty of introductory materials are available on polyhedral approaches in the context of combinatorial optimisation problems, such as [41, 55, 143]. I assume that the reader is familiar with the basics in convex analysis, linear programming (LP), especially flows in networks, graph theory, combinatorial optimisation, and an introductory knowledge of computational complexity theory. Chapters 1 and 2 in [37] can provide an easily accessible introduction to convex sets and convex polytopes. There are several classic books covering the areas mentioned above and dealing with combinatorial optimisation and computational complexity [44, 52, 108, 109, 154].

 Note *It may be fine to go straight to Chap. 3 after reading Sect. 2.5 and only return to the relevant sections of this chapter as required to refer to any of the definitions or notation for the precise meaning with which the terms are used in the book.*

---

[1] Henceforth I shall refer to this book as *Geometric Algorithms*.

2.1 Basic Notations

Let $\mathbb{R}$ denote the set of reals. Similarly, $\mathbb{Q}$, $\mathbb{Z}$, and $\mathbb{N}$ denote rationals, integers and natural numbers, respectively, and $\mathbb{B}$ stands for the binary set, $\{0, 1\}$. Let $\mathbb{R}_+$ denote the set of nonnegative reals. Similarly, the subscript $_+$ is understood with rationals, etc. Let $\mathbb{R}^d$ denote the set of d-tuples of reals. Similarly, the superscript ^{d} is understood with rationals, etc. Let $\mathbb{R}^{m \times n}$ denote the set of $m \times n$ real matrices. And a linear function $\mathbf{a} : \mathbb{R}^n \to \mathbb{R}$ is understood as

$$a(x) = a^t x = a_1 x_1 + \cdots + a_n x_n.$$

For x a real number, $|x|$ denotes the absolute value of x (that is, $|x| = x$ if $x \geq 0$ and $|x| = -x$ if $x < 0$). $\lfloor x \rfloor$ is the largest integer less than or equal to x, whereas $\lceil x \rceil$ is the smallest integer greater than or equal to x. For $n \geq 1$ and every integer a, $a \bmod n$ is the unique integer b in the set $\{0, \ldots, n - 1\}$ such that $\frac{(a-b)}{n}$ is an integer. We denote the empty set by $\emptyset$. The family of all subsets of the set S, including S itself and $\emptyset$ is the power set of S and is denoted by 2^S. $|S|$ is the cardinality (size) of the set S. The collection of k-element subsets of S is denoted by $\binom{S}{k}$. For any two sets S and T, let $S \triangle T$ denote the symmetric difference $(S \setminus T) \cup (T \setminus S)$ or equivalently $(S \cup T) \setminus (S \cap T)$.

2.2 Graph Theory

For basics in Graph Theory see the book by Bondy and Murty [125], or more recent ones by Béla Bollobás [36], or Reinhard Diestel [61].

> **Definition 2.1** (*Graph*) A (simple) graph $G = (V, E)$ consists of a finite set V of vertices and a family E of subsets of V of size two called edges; $E \subseteq \binom{V}{2}$. The order of G is $|V|$. The size of G is $|E|$. In addition, $|G|$ is used to denote $max(|V|, |E|)$. ♣

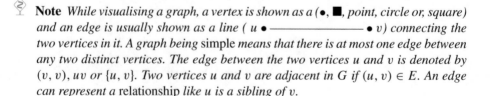

Note *While visualising a graph, a vertex is shown as a ($\bullet$, $\blacksquare$, point, circle or, square) and an edge is usually shown as a line (u $\bullet$ ——————— $\bullet$ v) connecting the two vertices in it. A graph being* simple *means that there is at most one edge between any two distinct vertices. The edge between the two vertices u and v is denoted by (v, v), uv or {u, v}. Two vertices u and v are adjacent in G if (u, v) $\in$ E. An edge can represent a* relationship *like u is a sibling of v.*

Definition 2.2 (*Degree of a Vertex, Neighbourhood*) Given a graph $G = (V, E)$, for $v \in V$, the neighbourhood of v is the set $N_G(v) = \{w \in V \mid \{v\} : vw \in E\}$. The degree of v is $\deg_G(v) = |N_G(v)|$. Similarly, for a set S of vertices in a graph G, let $N_G(S)$ denote the set of vertices having at least one neighbour in S. ♣

Note *If* $\deg_G(v) = 0$ *for all* $v \in V$ *then* G *is called an* empty graph. *If* $\deg_G(v) = k$ *for all* $v \in V$ *then* G *is called a* k-regular graph.

Definition 2.3 (*Subgraph, Induced Subgraph*) Given a graph $G = (V, E)$, we say that $G' = (V', E')$ is a subgraph of G if $V' \subset V$ and $E' \subset E$. Given $W \subseteq V$, the induced subgraph $G(W)$ of G on the vertex set W is the graph

$$G(W) = \left(W, E \cap \binom{W}{2} \right).$$

We denote by $G - W = G(V \setminus W)$ is the subgraph of G obtained by *deleting* the vertices in W and all edges whose both ends are in W. ♣

Definition 2.4 (*Directed Graph*) A directed graph is a graph $G = (V, E)$ in which every edge has a direction, that is, if $(i, j) \in E$ implies we have a direction $i \to j$. That is, $e = (u, v) \in E$ does not imply $(v, u) \in E$. ♣

Note *We use the term* arc *instead of edge for directed graphs. We will call a directed graph a* digraph. (*And some authors refer to it as* draph, *but I will not use this acronym.*)
Directed graphs are useful in applications where the implications are not both ways like u is the mother of v.

Definition 2.5 (*Bipartite Graph*) A graph $G = (V, E)$ is called a bipartite graph in case $V = V_1 \cup V_2$, $V_1 \cap V_2 = \emptyset$ and $E \subseteq \{(u, v) \mid u \in V_1, v \in V_2\}$. ♣

Note *Bipartite graphs play an important role in many applications and have interesting theoretical properties. In this book, we use bipartite flow problems a lot.*

Definition 2.6 (*Path*) A path is a graph P of the form

$$V(P) = \{v_0, v_1, \ldots, v_l\}, \quad E(P) = \{v_0 v_1, v_1 v_2, \ldots, v_{l-1} v_l\}$$

This path P is given by $v_0 v_1 \cdots v_l$. The vertices v_0 and v_l are the initial and terminal vertices of P respectively. And the length of P is defined as $l(P) = |E(P)|$. We say that P is a $v_0 - v_l$ path.

If path P is a subgraph of G, we say P is a path in G.

A Hamiltonian path in a graph is a path containing all the vertices of the graph.   ♣

 Note *In a path, the vertices v_i, $0 < i < l$, are distinct from each other and v_0. If not, we call the path, a* walk.

Definition 2.7 (*Cycle*) If a path $P = v_0 v_1 \cdots v_l$ is such that $l \geq 3$, $v_0 = v_l$, then P is said to be a cycle.   ♣

Definition 2.8 (*Hamiltonian Cycle*) A cycle containing all the vertices of a graph is said to be a Hamiltonian cycle of the graph.   ♣

In general, graph vertices can be members of any finite set. However, most graphs I deal with in this book are defined on subsets of positive integers. Therefore I introduce some specific notations.

Let n be an integer, $n \geq 3$. Let V_n be a set of *vertices*. Assuming, without loss of generality, that the vertices are numbered in some fixed order, we write $V_n = \{1, \ldots, n\}$. Equivalently, this set is denoted by $[n]$ in the literature. I prefer V_n as we are referring to the set of vertices. Let $E_n = \{(i, j) | i, j \in V_n, i < j\}$ be the set of *edges*. The cardinality of E_n is denoted by $p_n = n(n-1)/2$.

We denote the elements of E_n by e where $e = (i, j)$. We also use the notation ij for (i, j). Notice that, unlike the usual practice, an edge is assumed to be written with $i < j$.

Definition 2.9 (*Edge Label*) Let the elements of E_n be labelled as follows: $(i, j) \in E_n$, has the label, $l_{ij} = p_{j-1} + i$. ♣

This means, edges $(1, 2)$, $(1, 3)$, $(2, 3) \in E_3$ are labelled, 1, 2, and 3 respectively. Once the elements in E_{n-1} are labelled then the elements of $E_n \setminus E_{n-1}$ are labelled in increasing order of the first coordinate, namely i.

For a subset $F \subset E_n$ we write the *characteristic* vector of F by $x_F \in R^{P_n}$ where

$$x_F(e) = \begin{cases} 1 \text{ if } e \in F, \\ 0 \text{ otherwise.} \end{cases}$$

We assume that the edges in E_n are ordered in increasing order of the edge labels.

For a subset $S \subset V_n$ we write

$$E(S) = \{ij | ij \in E, i, j \in S\}.$$

Given $u \in R^{P_n}$, $F \subset E_n$, we define,

$$u(F) = \sum_{e \in F} u(e).$$

For any subset S of vertices of V_n, let $\delta(S)$ denote the set of edges in E_n with one end in S and the other in $S^c = V_n \setminus S$. For $S = \{i\}$, we write $\delta(\{i\}) = \delta(i)$.

Example 2.1 Consider the graph $G = (V, E)$, where $V = \{1, \ldots, 5\}$ and $E = \{(1, 5), (2, 3), (2, 5), (3, 4), (4, 5)\}$. We have $N_G(4) = \{3, 5\}$ and $\deg_G(5) = 3$.

Definition 2.10 (*Connected graph, Components*) A graph G is connected if for every pair, $\{x, y \in V | x \neq y\}$, there is a path from x to y. A connected component $C = (V(C), E(C))$ of a graph G is a maximal connected subgraph of G. $V = \cup_i V(C_i)$ and $E = \cup_i E(C_i)$, where, $C_i, i = 1, \ldots, k$, are the connected components of G. ♣

Note *A graph $G(V, E)$ has $|E| = |V| - 1$, then G has a single component and G is called a tree. We extend the definition of a component to digraphs using directed paths. We may just say a component instead of a connected component.*

Example 2.2 Consider $G = (V_5, E)$ with

$$E = \{(1, 2), (3, 4), (1, 5), (2, 5)\}.$$

Here G is not a connected graph. $\{1, 2, 5\}$ induces a maximal connected subgraph. Thus $C_1 = (\{1, 2, 5\}, \{(1, 2), (1, 5), (2, 5)\})$ is a connected component. And $\{3, 4\}$ induces the other component.

Definition 2.11 (*Diconnected Components*) [a]Given a digraph $G = (V, A)$, we say $u, v \in V$ are *diconnected* if there is a directed path from u to v and also there is a directed path from v to u in G. We write $u \leftrightarrow v$. $\leftrightarrow$ is an equivalence relation and it partitions V into equivalence classes, called diconnected components of G. Diconnected components are also called *strongly connected* components.   ♣

---

[a] Not to be confused with disconnected or bi-connected components.

Definition 2.12 (*Interface*) Given a digraph $G = (V, A)$, consider the diconnected components of G. An arc $e = (u, v)$ of G is called an *interface* if there exist two different diconnected components C_1, C_2 such that $u \in C_1$ and $v \in C_2$. The set of all interfaces of G is denoted by $I(G)$.   ♣

Definition 2.13 (*Bridge*) Given a graph $G = (V, E)$, an edge of G is called a *bridge* if $G - e$ has more components than G, where by component of a graph we mean a maximal connected subgraph of the graph.   ♣

Definition 2.14 (*Mixed Graph*) A *mixed graph* $G = (V, E \cup A)$ is such that it has both directed and undirected edges. A gives the set of directed edges (arcs). E gives the set of undirected edges(edges). If A is empty G is a graph and if E is empty G is a digraph.   ♣

In general, we call the elements of $E \cup A$, edges. Finding the diconnected components of a digraph can be achieved using a depth first search method in $O(|G|)$ where $|G|$ is given by $max\{|V|, |A|\}$. Similarly bridges in a graph can be found in $O(|G|)$.

A hypergraph generalises the concept of a graph.

Definition 2.15 (*Hypergraph*) A hypergraph $H = (V, E)$ is given by a finite set V, called the set of vertices, and a collection E of nonempty subsets of V, that is, $E \subset 2^V$, called hyperedges. If $|e| = k, \forall e \in E$ for some k, then we say H is $k - uniform$. $|V|$, and $|E|$ are called the order, and the size of the hypergraph, H respectively.   ♣

Note *Thus, any 2 − uniform hypergraph is a graph.*

Example 2.3 Consider $V = \{New \ Zealand, Australia, Tonga, England, France\}$. $E = \{\{New \ Zealand, Australia, Tonga\}, \{England, France\}\}$. $H = (V, E)$ is a hypergraph. And H is not uniform.

Definition 2.16 (*Matching*) Given a graph $G = (V, E)$, a matching M in G is a set of pairwise nonadjacent edges, that is, no two edges in this set share a common vertex. A *maximal* matching is a matching M of a graph G that is not a subset of any other matching, M'. A matching M is maximal if every edge in G has a non-empty intersection with at least one edge in M. A *maximum-cardinality* matching is a matching that contains the largest possible number of edges. A *perfect* matching is a matching that matches all vertices of the graph.   ♣

Remark (*Maximum vs Maximal*) A *maximum-cardinality* matching M, is such that $|M| \geq |M'|, \forall M' \in \mathcal{M}$, where $\mathcal{M}$ is the collection of all matchings in G. Thus the maximality of a matching need not imply maximum cardinality. A graph can contain a perfect matching only when the graph has an even number of vertices.

2.3 Convex Sets, Polytopes

Definition 2.17 Let $\mathbf{x}_1, \dots, \mathbf{x}_k \in \mathbb{R}^n$, and $\lambda = (\lambda_1, \dots, \lambda_k), \lambda_i \in \mathbb{R}, i = 1, \dots, k$. Then $\sum_{i=1}^{k} \lambda_i \mathbf{x}_i$ is called a *linear combination* of the vectors $\mathbf{x}_1, \dots, \mathbf{x}_k$.   ♣

Note *Depending on* λ, *we say,* $\sum_{i=1}^{k} \lambda_i \mathbf{x}_i$ *is a*

1. *conic combination, if* $\lambda_i \geq 0$,
2. *affine combination, if* $\sum_{i=1}^{k} \lambda_i = 1$, *and*
3. *convex combination, if it is conic and affine.*

We say a set of points $A = \{\mathbf{x}_1, \ldots, \mathbf{x}_k\} \subset \mathbb{R}^n$ is linearly dependent if there exist real numbers $\lambda_i \in \mathbb{R}$, $i = 1, \ldots, k$ such that

1. not all λ_i's are zero and
2. $\sum_{i=1}^{k} \lambda_i \mathbf{x}_i = 0$,

and A is said to be linearly independent if no such λ exists.

Definition 2.18 A set $C \subseteq \mathbb{R}^n$ is called a convex set if and only if $x, y \in C$ implies $\lambda x + (1 - \lambda) y \in C$, for all $0 \leq \lambda \leq 1$.   ♣

Definition 2.19 Given a convex set C, $x^* \in C$ is called an extreme point of C if and only if x^* is not a convex combination of any other $x, y \in C$. Or equivalently, Given a convex set C, $x^* \in C$ is called an extreme point of C if and only if $C \setminus \{x^*\}$ is convex.   ♣

Definition 2.20 Given a set V of m points $\mathbf{v}_1, \ldots, \mathbf{v}_m$ in $\mathbb{R}^n$, $\text{conv}(V) = \{x \mid x = \lambda_1 \mathbf{v}_1 + \cdots + \lambda_m \mathbf{v}_m, \sum_{1}^{m} \lambda_i = 1, \lambda_i \geq 0\}$, is called the convex hull of $\mathbf{v}_1, \ldots, \mathbf{v}_m$.   ♣

 Note *Similarly, linear, conic, and affine hulls are defined.*

Example 2.4 The following are obvious examples of convex sets:

- $\mathbb{R}^n, n \geq 1$,
- $\emptyset$, (in a vacuous sense)
- any open, closed or half-open interval in $\mathbb{R}$, and
- a halfspace, that is, $\{\mathbf{x} \in \mathbb{R}^n \mid a_1 x_1 + \cdots + a_n x_n \leq b \text{ for }, a_i, b \in \mathbb{R}\}$.

However, any finite set $S \subset \mathbb{R}^n$ with $|S| > 1$ is not a convex set. Thus, optimisation over such a set is a nonconvex optimisation problem.

Definition 2.21 (*Polytope, Dimension*) A convex polytope P in R^n is a compact (closed and bounded) convex set with a finite number (m) of extreme points $\mathbf{v}_1, \ldots, \mathbf{v}_m \in \mathbb{R}^n$. Dimension of a polytope P is $dim(P) = d$ in case d is the smallest number such that $P \subset R^d$. ♣

This definition leads to what is called a $\mathscr{V}$-presentation of a polytope.

Definition 2.22 A $\mathscr{V}$-presentation of a polytope P consists of integers n and m with $m \geq n \geqslant 1$, and m points $\mathbf{v}_1, \ldots, \mathbf{v}_m$ in $\mathbb{R}^n$ such that $P = \text{conv}(\{\mathbf{v}_1, \ldots, \mathbf{v}_m\})$. ♣

Definition 2.23 An $\mathscr{H}$-presentation of a polytope P consists of integers n and m with $m \geqslant n \geqslant 1$, a real $m \times n$ matrix A, and a vector $\mathbf{b} \in \mathbb{R}^m$ such that $P = \{x \in \mathbb{R}^n : Ax \leqslant b\}$. ♣

Note *Thus, a $\mathscr{V}$-presentation represents P as the convex hull of a finite set, and an $\mathscr{H}$-presentation represents P as an intersection of halfspaces. If every member $\mathbf{x}$ of a polytope P satisfies the inequalities $\mathbf{a_j}^T \mathbf{x} \leq b_j$ and $\mathbf{a_j}^T \mathbf{x} \geq b_j$, we say P lies on the hyperplane $\mathbf{a_j}^T \mathbf{x} = b_j$. If there are more than one such equality satisfied by all members of the polytope, we say the polytope lies on a flat.*
If the dimension of P, $dim(P)$ is less than n we will have at least one equality constraint satisfied by all $x \in P$ or P lying on a flat. We say P is not a full-dimensional polytope.

A *face* of a polytope P is P itself, the empty set, or the intersection of P with some supporting hyperplane. The 0-faces, 1-faces, and $(n - 1)$-faces of an n-polytope P are respectively, its vertices, edges, and facets. A $\mathscr{V}(\mathscr{H})$-presentation of P is irredundant if the omission of any of the points $\mathbf{v}_1, \ldots, \mathbf{v}_m$ (any of the inequalities in $Ax \leqslant b$) changes the polytope, reducing it in the first case and enlarging it in the second. In geometric terms, a $\mathscr{V}$-presentation is irredundant if each point $\mathbf{v}_i$ is a vertex of P, and when P is n-dimensional, an $\mathscr{H}$-presentation is irredundant if each inequality induces a facet of P.

A natural question:
Why do we need two such representations of a polytope? And a related question: Which representation is 'best'? Many optimisation problems turn out to be searching among a finite set of points. But the finite number could be very large and so computationally they turn out to be time-consuming to exhaust all of them. On the

other hand, a convex hull of finitely many points results in a convex polytope, and at least some optimisation problems over polytopes have easy algorithms. However, those methods might require a $\mathcal{H}$-presentation of the polytope to use the results from inequality systems. Thus both representations are needed. However, which presentation is preferred or 'best' depends on the particular problem situation. For instance,

[1] the convex hull of B^n, the unit hypercube has 2^n vertices but has the $\mathcal{H}$-presentation:

$$x_j \geq 0, \quad \& \quad x_j \leq 1, \forall\, j \in [n],$$

which has only $2n$ inequalities (facets);

[2] Consider a set S of $2n$ random points independently identically distributed uniformly in the unit hypercube. Now $conv(S)$ is expected to have exponentially many facets, the size of the $\mathcal{H}$-presentation, is approximately $\frac{2^n}{\sqrt{n\pi}}$. Thus, any such set $S, conv(S)$, has linear $\mathcal{V}$-presentation but exponential $\mathcal{H}$-presentation (See [96]); and

[3] A $n \times n$ square matrix $A = (a_{ij})$ is called doubly stochastic if all entries of the matrix are nonnegative and the sum of the elements in each row and each column is unity. The set of all doubly stochastic matrices can be given by, $\sum_j a_{ij} = 1, i \in [n]$, $\sum_i a_{ij} = 1, j \in [n]$, $a_{ij} \geq 0, i, j \in [n]$. This is a polytope and goes by the name Birkhoff polytope (see Sect. 2.5). Thus this polytope has a simple $\mathcal{H}$-presentation. However, Birkhoff–Von Naumann theorem, states that all $n \times n$ permutation matrices are vertices of this polytope. That is, the size of the $\mathcal{V}$-presentation of the polytope is exponential. But, linear optimisation over this polytope, with the additional requirement $a_{ij} \in \{0, 1\}$, that is, over exponentially many vertices of this polytope can be done efficiently.

Definition 2.24 (*Graph of a Polytope*) For a polytope P, the graph of P is defined as $G(P) = (V(P), E(P))$, where $V(P) = $ the set of 0-faces of P, and $E(P) = $ the set of 1-faces of P.   ♣

Graph-related definitions appear in Sect. 2.2

Definition 2.25 (*Diameter of P*) The distance between two vertices of a polytope P is defined as the minimum number of edges in a path joining them in $G(P)$ the graph of P. The diameter of a polytope is the greatest distance between two vertices of the polytope.   ♣

The diameter of a polytope is an interesting concept, especially in the context of expecting good adjacent vertex methods for optimisation over polytopes. The upper bound on the diameter of a d-polytope is given in [27] as at most $132^{d-3}(n - d + 52)$.

On the other hand, many polytopes of interest in combinatorial optimisation have small diameters. Padberg and Rao [137] show, for a certain class of polytopes, the diameter of any polytope is less than or equal to two. This class includes Birkhoff polytope on the one hand, and surprisingly, on the other hand, the polytope associated with the asymmetric travelling salesman problem polytope (see Sect. 2.6). Recently, a constructive proof is given by Rispoli, and Cosares [149] showing that the diameter of *STSP* polytope is at most 4, for every $n \geq 3$, and is thus independent of n. Since in our discussion of combinatorial optimisation using polytopes, we wish to go from a $\mathcal{V}$-presentation of a polytope, P to an irredundant $\mathcal{H}$-presentation (if possible), we need the definition of a *root* of a *face* of P [42]. Let V be the finite set of points defining P. For a face $F = P \cap \{x \mid ax = a_0\}$ of P, we call $x \in V \cap F$ a root of the inequality $ax \leq a_0$, and a root of F. Of prime interest for a polyhedron P is its minimal $\mathcal{H}$-presentation, $P = \{x \mid Ax \leq b, Bx = d\}$ where no equation from $Bx = d$ is implied by other equations of this system and the inequality system $Ax \leq b$ contains exactly one defining inequality for every facet of P.

In the next section, I go over a brief introduction to linear optimisation over polytopes.

2.4 Linear Programming

Linear Programming (LP) is one of the most frequently used models for solving optimisation problems directly, or indirectly in some subroutines used in solving such problems.

A typical LP is given in the form

$$\min \left\{ c^\top x : Ax = b, Bx \leq d, x \in \mathbb{R}^n \right\},$$

where A and B are a $m \times n$, and $k \times n$ real matrices respectively, $b \in \mathbb{R}^m$, and $d \in \mathbb{R}^k$, and $c \in \mathbb{R}^n$.

Notice that from the definition of a polyhedron we saw in the subsection on convex sets, in an LP, the objective function is minimised over the polyhedron $\mathcal{P} = \{x \in \mathbb{R}^n : Ax = b, Bx \leq d\}$ of feasible solutions.

Major algorithms have been proposed for solving LPs; they fall into three categories, namely [1] adjacent vertex methods like the simplex method, [2] interior point methods like that of Karmarkar's [100], and [3] ellipsoidal methods, like that of Khachiyan. Expositions on linear programming, for instance, Padberg's book [135] cover the three approaches, among them the Simplex method, invented by G. B. Dantzig stands out not only for its historical importance but also based on its performance in practice. Since I assume the reader is familiar with linear programming, here I very briefly touch on some milestones.

Adjacent Vertex Methods:

The simplex method starts with an initial feasible extreme point solution of the LP (if none exists, it stops, showing that the problem is infeasible), and in each step, it moves along an improving edge to an adjacent extreme point, until an optimal solution is found or it is established that the problem has no finite optimum. The algorithm works with a basis matrix corresponding to a vertex. Although the simplex method is adept to solve many practical problems, a major computational hurdle is that the pivot rule suggested by Dantzig, namely, moving to the most improving adjacent basis, could cycle among the different bases associated with the current vertex and come back to the same basic solution. This warranted new pivot rules that can avoid this. So other pivot rules emerged like Bland's rule that assured a remedy to cycling, ensuring finite convergence of the simplex algorithm. Since Victor Klee and George J. Minty [104] presented their example to show that the simplex method could take exponentially many steps before finding an optimum vertex, it is still unknown whether there is a pivot rule, that guarantees a polynomial upper bound on the number of steps performed by the simplex method. In the Klee–Minty example, the simplex method visits every extreme point. Klee–Minty polytope is expressed in terms of n inequality constraints on n non-negative variables. The number of extreme points is exponential in the size of the problem, 2^n. The Klee–Minty polytope is given by

$$x_1 \leq 5$$
$$4x_1 + x_2 \leq 25$$
$$8x_1 + 4x_2 + x_3 \leq 125$$
$$\vdots$$
$$2^n x_1 + 2^{n-1} x_2 + \cdots + 4x_{n-1} + x_n \leq 5^n$$
$$x_1 \geq 0, \ldots, x_n \geq 0$$

To maximise $x_1 + \cdots + x_n$, the simplex method, starting at $\mathbf{x} = \mathbf{0}$, goes through each of the extreme points before reaching the optimum solution at $(0, 0, \ldots, 0, 5^n)$.

'Does the decision version of the linear programming problem belong to P?' was settled by L. G. Khachiyan [103] in 1979. The clever idea to use Shor's [158] ellipsoid method, until then not used for a linear inequality system ($Ax \leq b$), especially, explicitly working with finite precision, Khachiyan showed that his method is polynomial in the input size. "This result has caused great excitement and stimulated a flood of technical papers" as noted by Robert G. Bland, Donald Goldfarb, and Michael J. Todd in their survey of ellipsoid method [35]. Some key questions one encounters frequently in convex analysis, got explored further and their interconnections, especially concerning computational complexity, became evident after this epoch-making discovery namely—linear programming problem is polynomially solvable, and the ellipsoidal method can be inventively used for solving an LP problem.

For instance, given a point y and a convex set K, is y a member of K? This question is called the membership problem. If $y \notin K$, finding a hyperplane separating y from K, goes by the name, the separation problem. Linear optimisation over K seeks a point in K that optimises the given linear function. If any of the three problems can be solved efficiently, all of them can be solved efficiently as demonstrated in [82] using the ellipsoidal method. Details of this work will be further explored in Chap. 7. On the other hand, the ellipsoidal approach as a polynomial alternative to the simplex method did poorly in real-world applications, but Narendra Karmarkar [100] led *interior point methods* perform as good as the simplex method and are now available in commercial as well as open source optimisation software alike (such as CPLEX [53], GUROBI [88], Julia [63] and the like).

Integer Linear optimisation:
LP problems with the additional restriction on the variables, requiring $x_j \in Z$, for $j \in J \subset [n]$, have challenged the researchers for long. Integral solutions only make sense in some applied problems where the variables can only take discrete values, and in theoretical problems like those involving graphs, or in computer science involving binary values, 0 or 1. We call them integer programming (*IP*) problems.

We say a linear inequality is *valid* for a polytope F if every member of F satisfies the inequality. In integer programming problems, we define the convex hull of integer feasible solutions in F the integer hull, $conv(F^*)$, where F^* is the set of integral extreme points of F.

Given an LP optimal solution y that is not integral, we wish to find a valid inequality that separates y from $conv(F^*)$, to impose that inequality as an additional requirement to be satisfied. This cute idea from Ralph Gomory [79] forms the basis for solving integer programming problems. Unfortunately proved impractical as the number of cuts required to obtain an integer solution exploded.

So the idea of finding a 'facet defining' cut which "cuts off" y along with other non-members as deep as permissible, that is, without cutting off any member of $conv(F^*)$ became fashionable. When successfully applied along with Branch-and-Bound strategies, this became very useful in practice. This results in Branch-and-Cut approaches, which are among the most successful approaches to solving difficult combinatorial optimisation problems, like the *TSP* solver Concorde for solving very large travelling salesman problem instances [4, 5]. We will not be needing these in what follows and so I refrain from further discussion of these approaches.[2] However, the well-solved 'network flow' problems are what I will be using repeatedly in studying pedigrees and designing algorithms.

2.5 Flows in Networks

The concepts and problems considered in this section are necessary for understanding the remaining Chapters. The maximum flow—minimal cut results by themselves form an important class of theorems in various other settings other than what is discussed. The concepts of flow and capacity are abstract and can be interpreted

---

[2] Chapter 8 discusses some of these in the context of a comparison of different methods for solving *STSP*.

differently in different real-world applications. A classic in the field is Ford and Fulkerson's *Flows in Networks*. Shrijver traces the history of the transportation and maximum flow problems in [153]. A later-day authoritative book came from Ahuja, Magnanti, and Orlin [1]. (Recent addition among others is Williamson's *Network Flow Algorithms* [163].)

Definition 2.26 (*Network*) A network N is given by $(G, s, t, \mathbf{c})$, where

- $G = (V, E)$ is a directed graph,
- $\mathbf{c} : E \to \mathbb{R}_+^{|E|}$, is a capacity function defined on the edges of G, and
- $s, t \in V$ are the source and sink respectively,

That is, $c_{i,j}$ gives the upper bound on the flow along the edge connecting the vertices i and j in the direction from i to j. We assume that if $\mathbf{c}$ is not given, the capacity is infinity.   ♣

Problem 2.1 (*Maximum flow problem*) Consider a network $N = (G = (V, E), s, t, \mathbf{c})$. Let $f : E \to \mathbb{R}_+^{|E|}$, denote the flow in the network, that is, for each edge $(i, j) \in E$ a flow, is given by $f_{ij} \geq 0$.

Find $f \in \mathbb{R}_+^{|E|}$ such that

1. the flow from s to t is a maximum,
2. for no arc the flow along the arc exceeds the capacity, [*Capacity restriction*] and
3. for every node in $v \in V, s \neq v \neq t$, the flow into the node v is equal to the flow out of the node. [*flow conservation*]

and this problem is called the maximum flow problem or *max-flow* problem.

 Note *Several practical problems yield themselves to be modelled as maximum flow problems [1]. In this book, flow problems appear in different contexts, from a theoretical and algorithmic perspective in Chaps. 4 through 6.*

Definition 2.27 (*Cut, Cut size, Cut capacity*) Suppose S is a set of vertices containing s but not containing t. Let $\bar{S}$ denote $V \setminus S$, the compliment of S. A cut $(S, \bar{S})$ is defined as the set of arcs, $e = (i, j) \in E$ such that, $i \in S \& j \in \bar{S}$. The size of a cut, $(S, \bar{S})$ is the number of edges from S to $\bar{S}$. The capacity of a cut $(S, \bar{S})$ is defined as $\sum_{(i,j)|i \in S, j \in \bar{S}} c_{i,j}$.   ♣

The set of *edge disjoint paths* is the collection of all paths from s to t such that no edge appears in more than one path.

Theorem 2.1 (Menger's Theorem) *A graph $G = (V, E)$ has k edge disjoint paths from s to $t \iff k$ is the size of the minimum directed $s - t$ cut.* ♡

Proof (See any book cited on Graph theory.)

From Menger's theorem, it is easy to see that one can have the flow from s to t along each of the edge-disjoint paths as much flow as possible only limited by the minimum of the capacities of the edges along any such path. Is that the maximum possible flow from s to t? In their celebrated work, entitled, *Flows in Networks*, Ford and Fulkerson [73] gave their intuitively appealing labelling algorithm for solving the maximum flow problem, but has a complexity of $O(|E| \cdot f^*)$, where f^* is the maximum flow in the network. The Ford–Fulkerson algorithm was eventually improved upon by Edmonds–Karp's algorithm, which does the same thing in $O(|V|^2 \cdot |E|)$ time, independent of the maximum flow value [68]. Notable other improved algorithms are available for solving the max-flow problem (see cited books on networks). Other variations of the maximum flow problem include minimum cost maximum flow problem and multicommodity flow problems among others. For these problems and developments in solving the maximum flow problem and its connection to the minimum cut problem, see cited books on combinatorial optimisation like [108]. Maximum flow problems are studied on hypergraphs as well (see Chap. 8 for citations). Flows in networks is a generic model to formulate other graph problems like bipartite matching, and operations research problems like the transportation problem defined below:

Problem 2.2 (*Transportation Problem*) Given m origins/sources O_i with availability a_i and n sinks/destinations D_j with demand b_j. In addition, we are given $d_{i,j} \in R_+$ the cost of transporting one unit from an origin O_i to destination D_j. Let $x_{i,j}$ denote amount transported from O_i to D_j. Find $X = (x_{i,j}) \in R^{m \times n}$, such that $\sum_i \sum_j d_{i,j} x_{i,j}$ is a minimum. If $\sum_i a_i = \sum_j b_j$ we say we have a *balanced* transportation problem. If we have in addition capacities $u_{i,j} \in R_+$ limiting the amount transported from O_i to D_j, we say we have a *capacited* transportation problem.

We are interested in a special case of the transportation problem and discuss the same in Sect. 2.5.1. And we need to work with a special type of network called layered networks as defined below:

Definition 2.28 (*Layered Network*) A network $N = (\mathcal{V}, \mathcal{A})$ is called a *layered network* if the node set of N can be partitioned into l sets $V_{[1]}, \ldots, V_{[l]}$ such that if $(u, v) \in \mathcal{A}$ then $u \in V_{[r]}, v \in V_{[r+1]}$ for some $r = 1, \ldots, l-1$. We say N has l layers. Nodes in $V_{[1]}, V_{[l]}$ are called sources and sinks respectively. ♣

In the case we are interested, we have *capacities both on arcs and nodes*. Any such problem can be recast as a flow problem with capacity restriction only on arcs, as observed by Ford and Fulkerson in [73].

2.5.1   FAT Problem, Rigid, Dummy Arcs

Consider a balanced transportation problem, in which some arcs called the *forbidden* arcs are not available for transportation. We call the problem of finding whether a feasible flow exists in such an incomplete bipartite network, a Forbidden Arcs Transportation (*FAT*) problem [127]. This could be viewed as a capacited transportation problem, as well. In general a *FAT* problem is given by $O = \{O_\alpha, \alpha = 1, \ldots, n_1\}$, the set of origins, with availability $a_\alpha \geq 0$ at O_α, $D = \{D_\beta, \beta = 1, \ldots, n_2\}$, the set of destinations, with a requirement $b_\beta \geq 0$ at D_β and $\mathcal{A} = \{(O_\alpha, D_\beta)|$ arc (O_α, D_β) is *not forbidden* $\}$, the set of arcs. We may also use (α, β) to denote an arc. Since the problem is balanced, we have, $\sum_\alpha a_\alpha = \sum_\beta b_\beta$. When any arc (α, β) has the capacity restriction, it is specified by $f_{\alpha\beta} \leq c_{\alpha\beta}$, as additional restriction. This problem is stated as Problem 2.3 and has many appearances in this book.

Problem 2.3 (*FAT-Problem*) Find f satisfying

$$\sum_{O_\alpha \in O} f_{\alpha\beta} = b(\beta), \text{ for all } D_\beta \in D \tag{2.1}$$

$$\sum_{D_\beta \in D} f_{\alpha\beta} = a(\alpha), \text{ for all } O_\alpha \in O \tag{2.2}$$

$$f_{\alpha\beta} \leq c_{\alpha\beta}, \text{ for all arc } (\alpha, \beta) \in \mathcal{A} \text{ with a capacity restriction} \tag{2.3}$$

$$f_{\alpha,\beta} \geq 0, \text{ for all arc}(\alpha, \beta) \in \mathcal{A} \tag{2.4}$$

Definition 2.29 (*Rigid Arcs*) Given a *FAT* problem with a feasible solution f we say $(\alpha, \beta) \in \mathcal{A}$ is a *rigid* arc in case $f_{\alpha,\beta}$ is the same in all feasible solutions to the problem. Rigid arcs have *frozen flow*.   ♣

Definition 2.30 (*Dummy Arc*) A rigid arc with zero frozen flow is called a *dummy* arc.   ♣

Fig. 2.1 *FAT* problem for
Example 2.5 [a]. Flows
shown along the arcs, $a(i)$
and $b(j)$ are shown in boxes,
for origin i and destination j

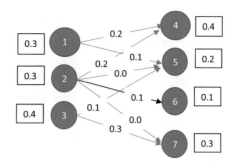

The set of rigid arcs in a *FAT* problem is denoted by $\mathcal{R}$. Identifying $\mathcal{R}$ is the *frozen flow finding* problem (*FFF* problem). Interest in this arises in various contexts (see [2]). Application in statistical data security is discussed by Gusfield [89]. The problem of protecting sensitive data in a two-way table, when the nonsensitive data and the marginal are made public is studied there. A sensitive cell is unprotected if its exact value can be identified by an adversary. This corresponds to finding rigid arcs and their frozen flows.

Example 2.5 [a] Consider the *FAT* problem given by $O = \{1, 2, 3\}$, with $a(1) = 0.3$, $a(2) = 0.3$, $a(3) = 0.4$, $D = \{4, 5, 6, 7\}$, with $b(4) = 0.4$, $b(5) = 0.2$, $b(6) = 0.1$, $b(7) = 0.3$. The set of forbidden arcs, $F = \{(1, 6), (1, 7), (3, 4), (3, 6)\}$. A feasible solution to this *FAT* problem is shown in Fig. 2.1.

Notice that, since $(2, 6)$ is the only arc entering 6 it is a rigid arc with frozen flow $f_{2,6} = 0.1$. Every other arc flow can be altered without losing feasibility, so they are all not rigid. Furthermore, there are no dummy arcs as the only rigid arc has a positive frozen flow.

[b] Consider the *FAT* problem discussed in [a] with the capacity restriction that $c_{1,4} = 0.2$ That is, $f_{1,4} \leq c_{1,4}$. Notice that f given in [a] is feasible for this problem as well. However, the set of rigid arcs will be different. As $f_{1,4} = c_{1,4}$, and in no feasible solution $f_{1,4}$ can be less than 0.2 as $b(4) = 0.4$ and $f_{2,4}$ can not be greater than 0.2. This is so because $f_{2,6} = 0.1$ in all feasible solutions to meet the demand at 6. Easy to check that all arcs are rigid in this case. Also, $(2, 5)$ and $(2, 7)$ are dummy arcs. And f is the unique feasible solution for this problem (see Fig. 2.2, where rigid and dummy arcs are shown with black and red arrows respectively).

Even though this problem can be posed as a linear programming problem we provide the graph algorithm developed in [89], in Appendix 2.7, with required definitions. We call this algorithm as *frozen flow finding* or *FFF* algorithm, and is a linear time algorithm, $O(|G_f|)$, where G_f is a bipartite graph defined on the node set of the *FAT* problem.

Fig. 2.2 *FAT* problem for
Example 2.5 [b]. Capacity
where present, flow are
shown along any arc; $a(i)$
and $b(j)$ are shown in boxes,
for origin i and destination j

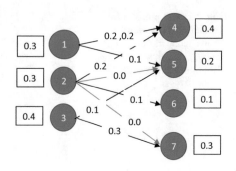

2.5.2 Two Partitions and a FAT Problem

Next, we give a *FAT* problem arising in the context of partitioning a finite set into
two different ways, which has applications in Sects. 4.4 and 5.4 of the book.

Definition 2.31 Suppose $\mathcal{D} \neq \emptyset$ is a finite set and $g : \mathcal{D} \to Q_+$, is a nonnegative function such that, $g(\emptyset) = 0$, for any subset S of $\mathcal{D}$, $g(S) = \sum_{d \in S} g(d)$.
Let $\mathcal{D}^1 = \{D^1_\alpha, \alpha = 1, \ldots, n_1\}$, $\mathcal{D}^2 = \{D^2_\beta, \beta = 1, \ldots, n_2\}$ be two non-empty
partitions of $\mathcal{D}$. (That is, $\bigcup_{\alpha=1}^{n_1} D^1_\alpha = \mathcal{D}$ and $D^1_s \bigcap D^1_r = \emptyset, r \neq s$. Similarly
$\mathcal{D}^2$ is understood.) ♣

Lemma 2.1 (Basic Lemma) *Given $D^i, i = 1, 2$, as given by Definition 2.31
above, consider the FAT problem defined as follows:*

*Let the origins correspond to D^1_α, with availability $a_\alpha = g(D^1_\alpha)$, $\alpha =
1, \ldots, n_1$ and the destinations correspond to D^2_β, with requirement $b_\beta =
g(D^2_\beta)$, $\beta = 1, \ldots, n_2$. Let the set of arcs be given by*

$$\mathcal{A} = \{(\alpha, \beta) | D^1_\alpha \cap D^2_\beta \neq \emptyset\}.$$

*Then $f_{\alpha,\beta} = g(D^1_\alpha \cap D^2_\beta) \geq 0$ is a feasible solution for the FAT problem con-
sidered.* ♡

Proof Since $\mathcal{D}^i, i = 1, 2$, are partitions the problem is balanced. It is easy to see that
$\sum_\alpha f_{\alpha,\beta} = g(D^2_\beta)$. That is, demands at every destination are met, without violating
any availability restrictions. Thus, f as defined is a feasible solution to the *FAT*
problem.

However, if there are capacity restrictions on the arcs then, Lemma 2.1 need not be true always. Several other *FAT* problems are defined and studied in the later sections of this book. *FAT* problems can be solved using any efficient bipartite maximal flow algorithm like Edmonds and Karp's algorithm (see [108]). If the maximal flow is equal to the maximum possible flow, namely $a(O)$, we have a feasible solution to the problem.

2.6 0/1-Polytopes

In this section, we bring together results on 0/1-polytopes that help us discuss and understand combinatorial optimisation problems. Günter Ziegler in his introduction to 0/1-polytopes [168] observes,

> A good grasp on the structure of 0/1-polytopes is important for the "polyhedral combinatorics" approach of combinatorial optimisation.

> This has motivated an extremely thorough study of some special classes of 0/1-polytopes such as the travelling salesman polytopes ... For example, [33] have proved the surprising fact that every 0/1-polytope appears as a face of a TSP-polytope.

Several operations research and economic problems can be formulated as mathematical programming problems with linear objectives seeking 0/1 integer solutions [60]. They necessitate the study of polytopes whose vertices are 0/1 vectors, called 0/1-polytopes. Problems involving graphs formulated as combinatorial optimisation problems (COP) also lead to the study of such polytopes [45] (see Chap. 1 for a formal definition of *COP* and some typical problems). State-of-the-art concepts, theoretical results, and algorithms for *COP* are presented in a recent book by Korte and Vygen [108] or the monumental work by Alexander Schrijver on combinatorial optimisation for a polyhedral combinatorics exposition exploring efficiency [154]. We encountered a well-known 0/1-polytope, namely, the Birkhoff Polytope earlier. The Birkhoff polytope is studied in different contexts, other than while studying doubly stochastic matrices, and is variously known as, (i) *assignment polytope*, or (ii) the bipartite matching polytope of the complete bipartite graph, $K_{n,n}$. This polytope recurs in research on other 0/1 polytopes, for instance, Louis J Billera and Aravamuthan Sarangarajan [33] and Manfred Padberg, and M. Rammohan Rao [137] study the adjacency structure of the assignment polytope to throw light on that of the asymmetric travelling salesman polytope. Studying the adjacency structure of 0/1-polytopes has interested researchers both from theoretical and practical algorithmic perspectives. The success of the simplex method, for solving linear programming problems, created interest in adjacent vertex improvement methods. This is one impetus for studying the adjacency structure of 0/1-polytopes [98]. It is well known that testing whether a given pair of vertices are adjacent can be done in polynomial time in some classical *COP* polytopes[3] like, matching polytopes [25, 45], set partitioning

---

[3] See cited articles/books for the description of the classical polytopes mentioned in this section.

polytopes [21–23] vertex packing polytopes [45] and, set packing polytopes [97]. On the other hand, Papadimitriou [138] has shown that the problem of checking nonadjacency on the travelling salesman polytope is **NP-complete**. Matsui shows that knapsack polytopes, set covering polytopes, among others, also share a similar fate [119].

Like testing nonadjacency on 0/1 polytopes, another interesting line of research is exploring the diameter of a polytope. In what follows we give some definitions and properties relating to polytopes, wherein F refers to the vertices of the polytope, $P = conv(F)$.

2.6.1   Some Properties of 0/1 Polytopes

Definition 2.32 (*Adjacency*) $\mathbf{x}, \mathbf{y} \in F$ are *adjacent* vertices of P = conv(F) if and only if, for every $\lambda, 1 > \lambda > 0$, $\lambda \mathbf{x} + (1 - \lambda) \mathbf{y}$ cannot be expressed as a convex combination of elements of $F \setminus \{\mathbf{x}, \mathbf{y}\}$. In other words, the line segment $[\mathbf{x}, \mathbf{y}]$ is an edge of the polytope, that is, it is a one-dimensional face of P. ♣

Similarly, we can define *nonadjacency* in polytopes. It is easy to observe that in Definition 2.32 of adjacency, if we are considering a 0/1-polytope, it is sufficient to consider convex combinations of vertices that agree with $\mathbf{x}$ and $\mathbf{y}$ on coordinates in which they themselves agree. We can have an equivalent definition of adjacency of $\mathbf{x}, \mathbf{y}$ in P as: any point $\mathbf{x}^0$ in the line segment $(\mathbf{x}, \mathbf{y})$ can be expressed as a convex combination of vertices of P in an unique manner. And so we have the easy-to-show equivalent definition of nonadjacency of vertices:

Definition 2.33 (*Nonadjacency*) $\mathbf{x}, \mathbf{y} \in F$ are *nonadjacent* in $conv(F)$, if and only if there exist a $S \subset F$ and a weight vector μ such that

- $S \cap \{\mathbf{x}, \mathbf{y}\} \neq \{\mathbf{x}, \mathbf{y}\}$,
- $\sum_{w \in S} \mu(w)\mathbf{w} = (\mathbf{x} + \mathbf{y})/2$, $\sum_{w \in S} \mu(w) = 1$, $\mu(w) > 0$, $\mathbf{w} \in S$.

Such a S is called a *witness for nonadjacency* of the given vertices, or *witness* for short. ♣

Given a polytope P we have defined earlier the *graph of P*, $G(P)$. $G(P)$ is also known as 1-skeleton of the polytope P.

A sequence $\rho = (\mathbf{x}^0, \mathbf{x}^1, \ldots, \mathbf{x}^K)$ of distinct vertices of P is called a *vertex sequence* (of P from $\mathbf{x}^0$ to $\mathbf{x}^K$). When a vertex sequence ρ contains $K + 1$ vertices, we say the length of ρ is K.

Definition 2.34 (*Monotone Vertex Sequence*) A vertex sequence $\rho = (\mathbf{x}^0, \mathbf{x}^1, \mathbf{x}^2, \ldots, \mathbf{x}^K)$ is called a *monotone vertex sequence* when it satisfies the condition that: for each index j, either

$$x_j^0 \le x_j^1 \le x_j^2 \le \cdots \le x_j^K \text{ or } x_j^0 \ge x_j^1 \ge x_j^2 \ge \cdots \ge x_j^K .$$

♣

Definition 2.35 (*Property A*) If two vertices $\mathbf{x}^1$ and $\mathbf{x}^2$ of P are not adjacent, then there exists a vertex $\mathbf{x}'$ of P such that $(\mathbf{x}^1, \mathbf{x}', \mathbf{x}^2)$ is a monotone vertex sequence of P.

♣

For 0/1-polytopes, using Definition 2.33, we have a vertex $\mathbf{y}$ in S which is different from both $\mathbf{x}^1$ and $\mathbf{x}^2$ and $\mathbf{y}$ agrees on coordinates in which the other two vertices themselves agree. It is easy to verify that $(\mathbf{x}^1, \mathbf{y}, \mathbf{x}^2)$ is a monotone vertex sequence. This fact appears as Lemma 2.1 in [120].

Definition 2.36 (*Property B*) If $(\mathbf{x}^1, \mathbf{x}^2, \mathbf{x}^3)$ is a monotone vertex sequence of P, then the vector $\mathbf{x}^1 - \mathbf{x}^2 + \mathbf{x}^3$ is a vertex of P.

♣

This property is not true for all 0/1 polytopes.

Denis Naddef and William Pulleyblank have defined the combinatorial property of polytopes, which is important in the study of pedigree polytopes, as I shall show in Chaps. 4 and 6.

Definition 2.37 (*Property Combinatorial*) If $\mathbf{x}^1$ and $\mathbf{x}^2$ are nonadjacent vertices of P then there exist two other vertices $\mathbf{y}^1$ and $\mathbf{y}^2$ of P such that $\mathbf{x}^1 + \mathbf{x}^2 = \mathbf{y}^1 + \mathbf{y}^2$. We say the vertex set F is a *combinatorial set*. The corresponding graph and polytope are called *combinatorial graph* and *combinatorial polytope* respectively.

♣

Definition 2.38 (*Hirsch*) Let P be any polytope in R^n. For any $\mathbf{c} \in R^n$ and for any vertex $\mathbf{x}^0$ of P, the following holds. If the problem of minimising $\mathbf{cx}$ over P has an optimum then there exists a vertex sequence $\rho = (\mathbf{x}^0, \mathbf{x}^1, \ldots, \mathbf{x}^K)$ of P satisfying

1. $\mathbf{x}^K$ is an optimal solution to the problem,
2. $\mathbf{x}^{i-1}$ and $\mathbf{x}^i$ are adjacent for all $i \in \{1, 2, \ldots, K\}$,
3. $K \leq f(P) - d(P)$ where K is the length of the sequence, $f(P)$ is number of facets of P and $d(P)$ is the dimension of P.   ♣

Definitions 2.37 and 2.38 are from [130]. See Sect. 2.3 for definitions of terms relating to polytopes. In addition to Hirsch property given by Definition 2.38, if we have $\mathbf{cx}^0 \geq \mathbf{cx}^1 \geq \cdots \geq \mathbf{cx}^K$ we have the *monotone* version of the Hirsch property. Denis Naddef [128] showed that 0/1-polytopes have Hirsch property. If a polytope satisfies the properties given by Definitions 2.35 and 2.36 then it has monotone Hirsch property as shown by Matsui and Tamura [120]. They also prove the monotone Hirsch property is true for all 0/1-polytopes.

Definition 2.39 (*Property P1*) If $\mathbf{x}^1$ and $\mathbf{x}^2$ are nonadjacent vertices of P then there exist other vertices $\mathbf{y}^1, \ldots, \mathbf{y}^r$ of P and positive integers $\lambda_1, \ldots, \lambda_r$ such that $\mathbf{x}^2 - \mathbf{x}^1 = \sum_{i=1}^{r} \lambda_i (\mathbf{y}^i - \mathbf{x}^1)$.   ♣

Definition 2.40 (*Strong Adjacency*) Let P be any polytope in R^n. Consider the problem of minimising $\mathbf{cx}$ over P, for $\mathbf{c} \in R^n$. If every best-valued (optimal) vertex of P is adjacent to some second best-valued vertex of P for each $\mathbf{c}$, we say P has the strong adjacency property.   ♣

The above two properties are as defined in [97].

Remark (*Implications*) We have the following implications among the properties discussed above:

- Every 0/1-polytope has properties Matsui and Tamara property A (Definition 2.35), and Hirsch property (Definition 2.38); and in fact 0/1-polytopes possess the monotone Hirsch property (see for proofs [120, 130] respectively).
- Matsui and Tamara properties A and B (Definitions 2.35 and 2.36) imply combinatorial property (Definition 2.37) (see [120]).
- Combinatorial property (Definition 2.37) implies property P1 (Definition 2.39), easily follows from the definitions.

- Property P1 (Definition 2.39) implies strong adjacency property (Definition 2.40) (proved in [97]).

Note *The properties discussed in this section can come in handy while design- ing algorithms to solve COP problems. In this book, the combinatorial property of pedigree polytopes will be used in Chaps. 4 and 6.*

2.6.2 New Results on Simplex Method for 0/1 LP

As we saw in Sect. 2.4, despite the excellent performance of the simplex method for solving linear programming application instances, it is still not known whether there exists a pivot rule with the guarantee of a polynomial number of adjacent basis changes. In a recent paper, Black et al. [34] study the behaviour of the simplex method for 0/1-linear programs. They present three pivot rules which require only a polynomial number of nondegenerate pivots to reach an optimal solution. One of their pivot rules is the True Steepest Edge pivot rule. And they prove that the True Steepest Edge pivot rule reaches an optimal solution within a strongly-polynomial number of nondegenerate steps. They interpret Edmonds and Karp's [68] max-flow algorithm, which we saw in Chap. 1, as moving along the steepest edges on the 1-skeleton of the 0/1 polytope of feasible flows. Similarly, they argue that computing a maximum matching using the shortest augmenting path algorithm corresponds to moving along the steepest edge direction on the 1-skeleton of the matching polytope. Our observation here is that avoiding non-determinism is the key to obtaining poly- nomially bounded algorithms for difficult problems, as well as using information on 1-skeleton of the polytope can be exploited while designing algorithms for linear optimisation over 0/1 polytopes.

2.7 Appendix on Frozen Flow Finding Algorithm

We give the following definitions in addition to the ones found in Sect. 2.5.1. Notice that G_f defined here is a mixed graph, while the residual graph G_f defined, in Chap. 1 has only directed arcs (see under Ford–Fulkerson method).

Definition 2.41 (G_f) For a feasible flow f for a *FAT* problem, we define a mixed graph $G_f = (\mathcal{V}, A \cup E)$ where $\mathcal{V}$ is as given in the *FAT* problem, and

$$A = \{(O_\alpha, D_\beta) | f_{\alpha,\beta} = 0, (\alpha, \beta) \in \mathcal{A}\} \cup \{(D_\beta, O_\alpha) | f_{\alpha,\beta} = c_{\alpha,\beta}, (\alpha, \beta) \in \mathcal{A}\},$$

$$E = \{(O_\alpha, D_\beta) | 0 < f_{\alpha,\beta} < c_{\alpha,\beta}, (\alpha, \beta) \in \mathcal{A}\}. \qquad \clubsuit$$

Definition 2.42 (*Flow change Cycle*) A simple cycle in G_f is called a *flow change cycle* (fc-cycle), if it is possible to trace it without violating the direction of any of the arcs in the cycle. Undirected edges of G_f can be oriented in one direction in one fc-cycle and the other direction in another fc-cycle.   ♣

Theorem 2.2 (Characterisation of Rigid Arcs [89]) *Given a feasible flow, f, to a FAT problem, an arc is rigid if and only if its corresponding edge is not contained in any fc-cycle in G_f.*   ♡

The proof of this is straightforward from the definitions given in this chapter. It is also proved in [89] that the set $\mathcal{R}$ (of rigid arcs + dummy arcs) is given by the algorithm[4] stated below:

Algorithm 2 FFF

Given: A Forbidden arcs transportation problem with a feasible flow f.
Find: The set of rigid arcs $\mathcal{R}$, in the bipartite graph of the problem.
    Construct The mixed graph G_f as per Definition 2.41.
    Find The diconnected components of G_f (say $C_1, \ldots, C_q$), as per Definition 2.11.
    Find The set of interfaces, $I(G_f)$, as per Definition 2.12.
    Find The set of all bridges $B(G_f)$ in the underlying graphs,
    treating each C_r as an undirected graph, as per Definition 2.13.
    Output $\mathcal{R} = I(G_f) \cup B(G_f)$. Stop.

We have a linear-time algorithm, as each of the steps can be done in $O(|G_f|)$.

---

[4] *FFF* stands for Frozen Flow Finding.

Chapter 3
Motivation for Studying Pedigrees

In Chap. 1, we had a glimpse of the standard formulation for solving the symmetric travelling salesman problem (*STSP*) given by Dantzig, Fulkerson and Johnson [56]. There are many formulations for this problem. We will discuss them in Chap. 8 while comparing their performances with the one given in this chapter, namely, the *MI*-formulation. This chapter is devoted to establishing the connection between *STSP* and our study of pedigrees and thus motivating why studying pedigree optimisation is important.

3.1 Notations and Definitions

Given $n > 3$, the number of cities, consider the least cost Hamiltonian cycle problem, a *COP* given by $(\mathbb{E}, F, c)$ (See Example 1.3.). We will henceforth refer to this problem as the symmetric travelling salesman problem (*STSP*). Recall that we have $d_{ij} = d_{ji}$, the distance from i to j, the reason for the adjective, 'symmetric'.

> **Definition 3.1** We say $t = (1, i_1, \ldots, i_{k-1}, 1)$ is a k-tour in case $(i_1, \ldots, i_{k-1})$ is a permutation of $(2, \ldots, k), k \leqslant n$. ♣

© The Author(s), under exclusive license to Springer Nature Singapore Pte Ltd. 2023
T. S. Arthanari, *Pedigree Polytopes*,
https://doi.org/10.1007/978-981-19-9952-9_3

Definition 3.2 The length of a k-tour is defined as $c(t)$ given by

$$c(t) = \sum_{r=1}^{k-2} d_{i_r i_{r+1}} + d_{1 i_1} + d_{i_{k-1} 1}.$$

♣

Definition 3.3 Let $\mathcal{T}_k$ denote the set of all k-tours and $\mathcal{T}_{ijk}$ denote the set of all k-tours in which is, i and j are adjacent.   ♣

Then we have $\mathcal{T}_k = \bigcup_{1 \leqslant i < j \leqslant k} \mathcal{T}_{ijk}$.

Definition 3.4 Let F_{ij}^k be a mapping from $\mathcal{T}_{ijk-1}$ to $\mathcal{T}_k$ such that for $t \in \mathcal{T}_{ijk}$, $F_{ij}^k(t)$ is the k-tour obtained from the $(k-1)$-tour t by inserting k between i and j.
  Let $c_{ijk} = d_{ik} + d_{jk} - d_{ij}$ for $4 \leqslant k \leqslant n, 1 \leqslant i, < j \leqslant k-1$.   ♣

Example 3.1 Let $n = 5$. We have the unique 3-tour $t = (1, 2, 3, 1) \in \mathcal{T}_3$. $F_{12}^4(t) = (1, 4, 2, 3, 1) \in \mathcal{T}_4$. Proceeding further, $F_{14}^5(t) = (1, 5, 4, 3, 2, 1)$, yielding a 5-tour. Consider $t = (1, 2, 3, 4, 1) \in \mathcal{T}_4$. Then t belongs to each one of $\mathcal{T}_{144}, \mathcal{T}_{344}, \mathcal{T}_{234}, \mathcal{T}_{124}$.

We now state the following results that can be easily proved.

Proposition 3.1 *Let $t_1, t_2 \in \mathcal{T}_{ijk-1}$. If $c(t_1) \leqslant c(t_2)$ then*

$$c(F_{ij}^k(t_1)) \leqslant c(F_{ij}^k(t_2)).$$

♠

Proposition 3.2 $\mathcal{T}_{k+1} = \bigcup_{1 \leqslant i < j \leqslant k} \{F_{ij}^{k+1}(t) \mid t \in \mathcal{T}_{ijk}\}$.   ♠

Proposition 3.3 $min_{t \in \mathcal{T}_{k+1}} c(t) = min_{1 \leqslant i < j \leqslant k} \{min_{t \in \mathcal{T}_{ijk}} c(t) + c_{ijk+1}\}$. ♠

Remark The symmetric travelling salesman problem can be seen as finding an optimal n-tour, given d_{ij}, $1 \leqslant i < j \leqslant n$, with $d_{ij} = d_{ji}$. Proposition 3.3 assures an optimal n-tour, if we have a subset of $(n - 1)$-tours which includes for each $1 \leqslant i < j \leqslant n - 1$, a $(n - 1)$-tour in which i and j are adjacent and it minimises the length of the tour among all such $(n - 1)$-tours in which i and j are adjacent. However, finding such $(n - 1)$-tours may not be an easy task.

Thus we really have a $(n - 3)$ stage decision problem, in which in stage $(k - 3)$, $4 \leqslant k \leqslant n$, we decide on where to insert k. In the beginning we have a 3-tour $(1, 2, 3, 1)$. In the first stage, we decide on where to insert 4 among the available edges $(1, 2)$, $(1, 3)$, *and* $(2, 3)$. Call this set U^3, that is, the set of available edges for 4 to be inserted. Depending on this decision we have a set of available edges for the second stage insertion.

In the second stage, we decide where to insert 5 among the available edges. For instance, if our decision in the first stage is to introduce 4 between i_4 and j_4. Then the available edges are

$$U^4 = \{(1, 2), (1, 3), (2, 3)\} \cup \{(i_4, 4), (j_4, 4)\} - \{(i_4, j_4)\}.$$

In general, U^{k-1} depends on the decisions made in stages preceding k, $4 \leqslant k \leqslant n$. We have

$$U^{k-1} = U^{k-2} \cup \{(i_{k-1}, k - 1), (j_{k-1}, k - 1,)\} - \{(i_{k-1}, j_{k-1})\}$$

for some $(i_{k-1}, j_{k-1}) \in U^{k-2}$. U^{k-1} gives the set of edges in the $(k - 1)$-tour, which results from the decisions made in the preceding stages.

The associated total cost of these decisions made at different stages until stage $n - 3$ is

$$c_{i_4 j_4 4} + c_{i_5 j_5 5} + \cdots + c_{i_n j_n n}.$$

We are interested in finding an optimal $(i_4, j_4), \ldots, (i_n, j_n)$ such that the total cost is minimum, thus producing a n-tour that solves the problem. The length of this tour is given by $(d_{12} + d_{13} + d_{23}) + \sum_{k=4}^{n} c_{i_k j_k k}$. Here $(d_{12} + d_{13} + d_{23})$ is the length of the initial 3-tour which is independent of the decisions subsequently made.

3.2 A 0/1 Programming Formulation of the STSP: MI-Formulation

In this section, we describe a $0/1$ integer programming formulation of *STSP* known as the *MI*-formulation [20]. Let

$$x_{ijk} = \begin{cases} 1 & \text{if in stage } (k-3) \text{ the decision is to insert } k \text{ between } i \text{ and } j, \\ & 1 \leqslant i < j \leqslant k-1, \\ 0 & \text{otherwise.} \end{cases}$$

Assume that the indices ijk are ordered for each k according to the edge label of (i, j). Let the number of such indices be denoted by $\tau_n = \sum_4^n \frac{(k-1)(k-2)}{2}$. Recall B^d denotes $\{0, 1\}^d$.

Definition 3.5 (*Feasible Decision Vector*) Given $X = \left(x_{124}, \ldots, x_{n-2,n-1,n}\right) \in B^{\tau_n}$. Let $\left(x_{124}, \ldots, x_{k-3,k-2,k-1}\right)$, be denoted by $X/k - 1$.
  We say X is a feasible decision vector in case,
  (i) For every $k = 4, \ldots, n$

$$\sum_{1 \leqslant i < j \leqslant k-1} x_{ijk} = 1, \tag{3.1}$$

that is, k is inserted between i and j for exactly one edge (i, j); and
  (ii) $x_{ijk} = 1$ implies $T_{k-1}(X) \in \mathscr{T}_{ijk-1}$, where $T_{k-1}(X)$ is the $(k-1)$-tour resulting from the preceding decisions, $X/k - 1$.
  In other words, X is a feasible decision vector if $x_{i_k j_k k} = 1$ implies $(i_k, j_k) \in U^{k-1}, 4 \leqslant k \leqslant n$.   ♣

Example 3.2 For $n = 6$, let $x_{124} = 1$, $x_{145} = 1 \& x_{236} = 1$, then $X = (100; 000100; 0010000000)$ is a feasible decision vector as

$$\sum_{1 \leqslant i < j \leqslant k-1} x_{ijk} = 1 \quad \text{for } k = 4, 5, \text{ and } 6 \tag{3.2}$$

and $x_{124} = 1$, requires $T_3(X) \in \mathscr{T}_{123}$. And this is true as

$$T_3(X) = (1, 2, 3, 1).$$

  Similarly $x_{145} = 1 \Rightarrow T_4(X) = (1, 4, 2, 3, 1) \in \mathscr{T}_{144}$ and $x_{236} = 1 \Rightarrow T_5(X) = (1, 5, 4, 2, 3, 1) \in \mathscr{T}_{235}$.

However, $X = (100, 100000, 0010000000)$ is not a feasible decision vector, because $x_{125} = 1$ requires $T_4(X)$ to be in $\mathcal{T}_{124}$. But $(1, 4, 2, 3, 1)$, the resulting 4-tour from the earlier decision $x_{124} = 1$, does not satisfy this requirement.

Let $\mathfrak{J}$ be the set of all feasible decision vectors. We can state the multistage decision process as

Problem 3.1 Find $X^* \in \mathfrak{J}$ such that $c(X^*) = \min_{X \in \mathfrak{J}} c(X)$ where

$$c(X) = \sum_{k=4}^{n} \sum_{1 \leqslant i < j \leqslant k-1} c_{ijk} x_{ijk}.$$

This problem is a combinatorial optimisation problem corresponding to the multistage decision problem we have constructed. The task ahead is discovering more about the structure of this problem.

We shall now show, how $X \in \mathfrak{J}$ can be characterised by a set of linear equalities and inequalities along with the requirement $X \in B^{\tau_n}$. Notice that we already have for any $X \in \mathfrak{J}$

$$\sum_{1 \leqslant i < j \leqslant k-1} x_{ijk} = 1 \quad \text{for, } 4 \leq k \leq n. \tag{3.3}$$

In addition, x_{ijk} cannot be 1 if $(i, j) \notin T_{k-1}(X)$.

Condition (ii) of Definition 3.5 states that $x_{ijk} = 1 \Rightarrow (i_k, j_k) \in U^{k-1}; 4 \leqslant k \leqslant n$. We express this as a linear inequality as follows:

For all X, we have $(1, 2), (1, 3)$ & $(2, 3) \in T_3(X)$ as the initial 3-tour is $(1, 2, 3, 1)$. And any edge (i, j) among these edges is available in all sets $U^{k-1}, 4 \leqslant k \leqslant n$ unless $x_{ijk} = 1$ for some k. Since we begin with the 3-tour and at most one of the $x_{ijk} = 1, 4 \leqslant k \leqslant n$ for each $(i, j); 1 \leqslant i < j \leqslant 3$ we have the following constraint

$$\sum_{4 \leqslant k \leqslant n} x_{ijk} \leqslant 1. \tag{3.4}$$

Now consider other $(i, j)'s$, for $4 \leqslant j \leqslant n - 1$ and $1 \leqslant i < j$. x_{ijk} cannot be 1 unless (i, j) is an edge in the $(k-1)$-tour resulting from earlier decisions given by $X/k - 1$. However, (i, j) is created / *generated* only in one of the two ways, given below:

Either (i) $x_{rij} = 1$ for some $1 \leqslant r < i$ or (ii) $x_{isj} = 1$ for some $i + 1 \leqslant s < j$. Therefore, if

$$\sum_{1 \leqslant r \leqslant i-1} x_{rij} + \sum_{i+1 \leqslant s \leqslant j-1} x_{isj} = 1, \tag{3.5}$$

then the edge (i, j) is present at the kth stage and hence x_{ijk} can either be 0 or 1 for any $k \geqslant j + 1$.

Suppose

$$\sum_{1\leqslant r\leqslant i-1} x_{rij} + \sum_{i+1\leqslant s\leqslant j-1} x_{isj} = 0, \tag{3.6}$$

then the edge $(i,\,j)$ is not available for insertion for any $k \geqslant j+1$ and so $\sum_{j+1\leqslant k\leqslant n} x_{ijk} = 0$. Hence we have

$$\sum_{j+1\leqslant k\leqslant n} x_{ijk} \leqslant \sum_{1\leqslant r\leqslant i-1} x_{rij} + \sum_{i+1\leqslant s\leqslant j-1} x_{isj}$$

$$\Rightarrow -\sum_{1\leqslant r\leqslant i-1} x_{rij} - \sum_{i+1\leqslant s\leqslant j-1} x_{isj} + \sum_{j+1\leqslant k\leqslant n} x_{ijk} \leqslant 0. \tag{3.7}$$

Now Problem 3.1 can be given a 0/1 programming formulation as Problem 3.2.

Problem 3.2

minimise $\displaystyle\sum_{k=4}^{n} \sum_{1\leqslant i<j\leqslant k-1} c_{ijk} x_{ijk}$

subject to $\displaystyle\sum_{1\leqslant i<j\leqslant k-1} x_{ijk}$ $\qquad\qquad = 1, \quad 4 \leqslant k \leqslant n \tag{3.8}$

$\displaystyle\sum_{k=4}^{n} x_{ijk}$ $\qquad\qquad\qquad \leqslant 1, \quad 1 \leqslant i < j \leqslant 3, \tag{3.9}$

$\displaystyle -\sum_{r=1}^{i-1} x_{rij} - \sum_{s=i+1}^{j-1} x_{isj} + \sum_{k=j+1}^{n} x_{ijk} \;\leqslant\; 0, \quad 4 \leqslant j \leqslant n-1, \quad 1 \leqslant i < j,$
$$\tag{3.10}$$

$\qquad\qquad\qquad\qquad x_{ijk} \qquad\qquad = 0 \text{ or } 1, \quad 1 \leqslant i < j < k, \quad 4 \leqslant k \leqslant n.$
$$\tag{3.11}$$

Remark The objective function is the same as that of Problem 3.1.

Let $E_{[n]}$ denote the matrix corresponding to the constraints given by (3.8). $E_{[n]}$ is a $(n-3) \times \tau_n$ matrix of the following form:

$$E_{[n]} = \begin{bmatrix} e_{\frac{3\times 2}{2}} & 0 \cdot\cdot & & 0 \\ 0 & \cdots & & 0 \\ \cdot & \cdots & & \cdot \\ 0 & \cdot\cdot\, 0 & & e_{\frac{(n-1)(n-2)}{2}} \end{bmatrix},$$

where e_k is a vector each of whose coordinates is 1.

Consider the matrix of coefficients corresponding to constraint set (3.8)–(3.10). Add the following constraints that are always satisfied as x_{ijk} are non-negative:

$$-\sum_{r=1}^{i-1} x_{rin} - \sum_{s=i+1}^{n-1} x_{isn} \leqslant 0, \quad i = 1, \ldots, n-1, \tag{3.12}$$

Relax the integer restriction on X. Let $A_{[n]}$ be the matrix corresponding to constraints (3.9)–(3.12) without constraint (3.11). Let u_{ij} be the slack variable added to the constraint corresponding to (i, j). Let $U = [u_{ij}]$ be the vector of slack variable.

We get the following problem.

Problem 3.3

$$\min \quad c'X$$

$$\text{s.t.} \quad \begin{bmatrix} E_{[n]} & \mathbf{0} \\ A_{[n]} & I \end{bmatrix} \begin{bmatrix} X \\ U \end{bmatrix} = \begin{bmatrix} e_{n-3} \\ e_3 \\ 0 \end{bmatrix},$$

$$X, U \geqslant 0. \tag{3.13}$$

Note that adding the redundant constraints (3.12) is done to bring out the connection between the slack variables of Problem 3.3 and the edge-tour incident vectors of n-tours given by integer X feasible to Problem 3.3.

Theorem 3.1 *Any integer feasible solution to Problem 3.3 is a basic solution and has the following property.*

Let the submatrix of $A_{[n]}$ corresponding to the columns of $x_{i_k j_k k} = 1$, $k = 4, \ldots, n$ be denoted by Q. Then any row of Q is such that either

(i) all columns in a row are zeroes, or

(ii) exactly one of the elements is $+1$ and the rest are zeroes in the row, or

(iii) there is one -1 and one $+1$ in the row and the rest are zeroes, or

(iv) there is one -1 in the row and the rest are zeroes.

Moreover, any such solution corresponds to an n-tour. ♡

Proof Consider the square matrix B obtained by taking the columns corresponding to $x_{i_k j_k k} = 1, 4 \leqslant k \leqslant n$ and the columns corresponding to the slack variables u_{ij}. We have

$$B = \begin{bmatrix} I & 0 \\ Q & I \end{bmatrix},$$

where Q is the submatrix of $A_{[n]}$ corresponding to the columns $x_{i_k j_k k} = 1$, $4 \leqslant k \leqslant n$.

$$B^{-1} = \begin{bmatrix} I & 0 \\ -Q & I \end{bmatrix}.$$

Let Q_{ij} denote the row corresponding to the edge (i, j).

Case (i): $1 \leqslant i < j \leqslant 3$.

In this case, either no x_{ijk} is positive for edge (i, j) or at most one of them is equal to 1 in any integer feasible solution. This implies either

(a) Q_{ij} is a zero vector where we have an instance of (i) or (b) Q_{ij} has a single 1 and rest zeroes, where we have an instance of (ii). In these rows, there can be no -1's.

Case (ii): $1 \leqslant i < j, 4 \leqslant j \leqslant n - 1$.

Using the fact that for any (i, j) at most one of the x_{ijk} can be equal to 1 in any integer feasible solution to the problem, there can be at most one $+1$ in any of these rows.

However, this $+1$ cannot occur without a -1 in the same row since

$$- \sum_{r=1}^{i-1} x_{rij} - \sum_{s=i+1}^{j-1} x_{isj} + \sum_{k=j+1}^{n} x_{ijk} \leqslant 0 \tag{3.14}$$

If all x_{rij} or $x_{isj} = 0$ then $\sum_{k=j+1}^{n} x_{ijk} = 1$; and so cannot satisfy this constraint (3.14). Therefore, at least one of the x_{rij} or $x_{isj} = 1$. But for any r at most one $x_{ijr} = 1$. Thus, there is exactly one -1 in row Q_{ij}. This leads to an instance of (iii). On the other hand, if x_{rij}, as well as x_{isj}, are zeroes then x_{ijk} must all be zeroes. We have an instance of (i). Finally if one of the x_{rij} or $x_{isj} = 1$ and for all k $x_{ijk} = 0$, we have an instance of (iv).

Now we prove that any such solution corresponds to a tour. Consider $x_{i_k j_k k} = 1, 4 \leqslant k \leqslant n$. Insert in the 3-tour $(1, 2, 3, 1)$, city 4 between (i_4, j_4) and obtain a 4-tour. Assume introducing $5, \ldots, k$ in this manner in the 4-tour, $\ldots, (k - 1)$-tour respectively we obtain a k-tour. We shall show that introducing $(k + 1)$ in the unique k-tour obtained will result in a $(k + 1)$-tour.

We need to show that (i_{k+1}, j_{k+1}), for $1 \leqslant i_{k+1} < j_{k+1} \leqslant k$ is an edge available in

$$U^k \stackrel{\text{def}}{=} \{(1, 2), (1, 3), (2, 3)\} \cup_{r=4}^{k} \{(i_r, r), (j_r, r)\} - \cup_{r=4}^{k} \{(i_r, j_r)\},$$

If $(i_{k+1}, j_{k+1}) \notin U^k$, then it must be either

(a) (i_{k+1}, j_{k+1}) is (i_r, j_r) for some $4 \leqslant r \leqslant k$ or

(b) (i_{k+1}, j_{k+1}) is (i, r) with $i_r \neq i \neq j_r, 4 \leqslant r \leqslant k$.

However, (a) cannot happen as for any edge (i, j), $x_{ijk} = 1$ for at most one k and already $x_{i_r, j_r, r} = 1, 4 \leqslant r \leqslant k$.

If (b) happens, then the constraint corresponding to (i, r) will be violated and X cannot be feasible for the problem. This leads to a contradiction. Hence $(i_{k+1}, j_{k+1}) \in U^k$, thus any such solution corresponds to a tour.

Note *Any n-tour corresponds to an integer solution to Problem 3.3. Thus, there is a 1 − 1 correspondence between n-tours and the integer feasible solutions to Problem 3.3. Notice that U^k is the set of edges in the k-tour, resulting from the insertion decisions made, as per X, inserting 4, ..., k.*

3.3 Properties of MI-Formulation

In this section, we explore the properties of the new formulation.

Lemma 3.1 *Let U denote the vector of slack variables in Problem 3.3. Let (X, U) be any integer feasible solution to Problem 3.3. Then U is the edge-tour incidence vector of the n-tour given by (X, U), that is*

$$u_{ij} = \begin{cases} 1 & \text{if edge } (i, j) \text{ is present in the n-tour,} \\ 0 & \text{otherwise.} \end{cases}$$

♡

Proof Consider $1 \leqslant i < j \leqslant 3$ then from Eq. (3.9) we have

$$u_{ij} = 1 - \sum_{k=4}^{n} x_{ijk},$$

$u_{ij} = 0 \Rightarrow \sum_{k=4}^{n} x_{ijk} = 1$, which implies that for some $4 \leqslant k \leqslant n$, $x_{ijk} = 1$, i.e, (i, j) is not in the n-tour.

Conversely, suppose (i, j) is not in the n-tour, then we have $x_{ijk} = 1$ for some k, $4 \leqslant k \leqslant n$ which implies that $u_{ij} = 0$.

Now consider $1 \leqslant i < j$, $4 \leqslant j \leqslant n - 1$,

$$u_{ij} = \sum_{1 \leqslant r \leqslant i-1} x_{rij} + \sum_{i+1 \leqslant s \leqslant j-1} x_{isj} - \sum_{j+1 \leqslant k \leqslant n} x_{ijk},$$

(i, j) is not present in the n-tour if

$$\sum_{1 \leqslant r \leqslant i-1} x_{rij} + \sum_{i+1 \leqslant s \leqslant j-1} x_{isj} = 0 \Rightarrow \sum_{j+1 \leqslant k \leqslant n} x_{ijk} = 0,$$

$$\text{or} \sum_{1 \leqslant r \leqslant i-1} x_{rij} + \sum_{i+1 \leqslant s \leqslant j-1} x_{isj} = 1 \text{ and } \sum_{j+1 \leqslant k \leqslant n} x_{ijk} = 1.$$

Hence, $u_{ij} = 0$ if (i, j) is not present in the n-tour. Conversely, if $u_{ij} = 0$ we show that (i, j) is not present in the n-tour.

$$u_{ij} = 0 \Rightarrow \sum_{1 \leqslant r \leqslant i-1} x_{rij} + \sum_{i+1 \leqslant s \leqslant j-1} x_{isj} - \sum_{j+1 \leqslant k \leqslant n} x_{ijk} = 0$$

$$\Rightarrow \sum_{1 \leqslant r \leqslant i-1} x_{rij} + \sum_{i+1 \leqslant s \leqslant j-1} x_{isj} = \sum_{j+1 \leqslant k \leqslant n} x_{ijk}.$$

There are two cases:

(a)

$$\sum_{1 \leqslant r \leqslant i-1} x_{rij} + \sum_{i+1 \leqslant s \leqslant j-1} x_{isj} = 0.$$

This implies that edge (i, j) is not created and hence is not available for insertion of any k, $j + 1 \leqslant k \leqslant n$. Hence, (i, j) is not in the n-tour.

(b)

$$\sum_{1 \leqslant r \leqslant i-1} x_{rij} + \sum_{i+1 \leqslant s \leqslant j-1} x_{isj} = 1 = \sum_{j+1 \leqslant k \leqslant n} x_{ijk},$$

which implies that edge (i, j) is created, but then some k, $j + 1 \leqslant k \leqslant n$ is inserted between (i, j). Hence, (i, j) is not in the n-tour.

Hence, $u_{ij} = 0$ if (i, j) is not in the n-tour.

Lemma 3.2 *Corresponding to any feasible solution to Problem 3.3, we have*

(i) $\sum_{1 \leqslant i < j \leqslant n} u_{ij} = n$, and
(ii) $\forall 1 \leqslant i < j \leqslant n, 0 \leqslant u_{ij} \leqslant 1$.   ♡

Proof We shall show part (i) is true for any feasible solution (X, U) to Problem 3.3. As (X, U) is feasible, we have

$$E_{[n]}X = e_{n-3}, \tag{3.15}$$

$$A_{[n]}X + IU = \begin{bmatrix} e_3 \\ 0 \end{bmatrix}. \tag{3.16}$$

Now sum the $n(n - 1)/2$ rows of (3.16). We get

$$-\sum_{k=4}^{n} \sum_{1 \leqslant i < j \leqslant k-1} x_{ijk} + \sum_{1 \leqslant i < j \leqslant n} u_{ij} = 3. \tag{3.17}$$

But, $\sum_{k=4}^{n} \sum_{1 \leq i < j \leq k-1} x_{ijk} = n - 3$ as obtained from the sum of the $(n - 3)$ rows of (3.15).

Proof of part (ii) follows easily.

Observe that $c' = \left(c_{124}, \ldots, c_{12n}, \ldots, c_{(n-2)(n-1)n}\right)$ is such that $c' = -d' A_{[n]}$. Consider any solution (X, U) to Problem 3.3. Then

$$A_{[n]} X + I U = \begin{bmatrix} e_3 \\ 0 \end{bmatrix}.$$

Premultiply both sides by d'. Now

$$d' U = d' \begin{bmatrix} e_3 \\ 0 \end{bmatrix} - d' A_{[n]} X$$

$$= d' \begin{bmatrix} e_3 \\ 0 \end{bmatrix} + c' X. \tag{3.18}$$

But $d' \begin{bmatrix} e_3 \\ 0 \end{bmatrix} = d_{12} + d_{13} + d_{23}$ is a constant. Therefore, minimising $d' U$ is equivalent to minimising $c' X$.

Thus, we have Problem 3.4 which is equivalent to Problem 3.3.

Problem 3.4

$$\text{minimise} \quad d' U \quad \text{such that} \quad \begin{bmatrix} E_{[n]} & \mathbf{0} \\ A_{[n]} & I \end{bmatrix} \begin{bmatrix} X \\ U \end{bmatrix} = \begin{bmatrix} e_{n-3} \\ e_3 \\ 0 \end{bmatrix}, \tag{3.19}$$

$$X, U \geqslant 0.$$

Remark Any n-tour corresponds to an integer basic feasible solution. However, there are basic feasible solutions that are non-integer as illustrated in the following example.

Example 3.3 Let $x_{124} = x_{134} = x_{135} = x_{245} = \frac{1}{2}$. There is a basic feasible solution to Problem 3.4 with corresponding $u_{12} = \frac{1}{2}, u_{13} = 0, u_{23} = 1, u_{14} = 1, u_{24} = 0, u_{34} = \frac{1}{2}, u_{15} = u_{25} = u_{35} = u_{45} = \frac{1}{2}$.

Let

$$\zeta(n) = \left\{ X \mid E_{[n]} X = e_{n-3}, X \geqslant 0 \right\}, \tag{3.20}$$

$$\mathscr{U}(n) = \left\{ U \mid U = \begin{bmatrix} e_3 \\ 0 \end{bmatrix} - A_{[n]} X \geqslant 0, X \in \zeta(n) \right\}. \tag{3.21}$$

Lemma 3.3 *Let*

$$U^* = \begin{bmatrix} e_3 \\ 0 \end{bmatrix} - A_{[n]}X^* \geqslant 0$$

for any integer $X^* \in \zeta(n)$. Then U^* is an extreme point of $\mathscr{U}(n)$. ♡

Proof Let $U, V \in \mathscr{U}(n) - \{U^*\}$. We shall show that $\lambda U + (1 - \lambda)V$ for $\lambda \in (0, 1)$ belongs to $\mathscr{U}(n) - \{U^*\}$.

We have $U \neq U^* \neq V$

As $U, V \in \mathscr{U}(n) - \{U^*\}$ there exists X, Y such that

$$U = \begin{bmatrix} e_3 \\ 0 \end{bmatrix} - A_{[n]}X \geqslant 0$$

and

$$V = \begin{bmatrix} e_3 \\ 0 \end{bmatrix} - A_{[n]}Y \geqslant 0.$$

Note that $X \neq X^* \neq Y$.

Now $\lambda U + (1 - \lambda)V \in \mathscr{U}(n)$ since $\mathscr{U}(n)$ is a convex set. We want to prove that $\lambda U + (1 - \lambda)V \in \mathscr{U}(n) - \{U^*\}$, as well.

Suppose this is not true. Then $\lambda U + (1 - \lambda)V = U^*$. Since X^* is integral U^* is also integral. We know that

$$\sum_{1 \leqslant p < q \leqslant n} U^*_{pq} = \sum_{1 \leqslant p < q \leqslant n} U_{pq} = \sum_{1 \leqslant p < q \leqslant n} V_{pq} = n$$

as these correspond to feasible solutions to Problem 3.3. In addition, note that for any feasible solution (X, U) to Problem 3.3, $0 \leqslant u_{ij} \leqslant 1$. Therefore if any coordinate of U^* is zero the corresponding coordinates of U as well as V have to be zero, as λ, $(1 - \lambda) > 0$, and $U, V \geqslant 0$.

Thus $U = V = U^*$, which leads to a contradiction as $U, V \in \mathscr{U}(n) - \{U^*\}$. Hence the result.

Note *The fact used in the proof, about 0/1 vectors and their convex combinations, is common knowledge in combinatorial optimisation.*

$\mathscr{U}(n)$ is the orthogonal projection of the polytope $\zeta(n)$. It is expected that some of the projected extreme points are no longer extremal in the projection as shown in the following two examples.

Example 3.4 Consider the fractional basic feasible solution given earlier for the 5-city problem in Example 3.3. This solution can be written as a convex combination of the solutions given by

(a) $x_{134} = 1$, $x_{125} = 1$, $u_{23} = u_{14} = u_{34} = u_{15} = u_{25} = 1$, and

(b) $x_{134} = x_{345} = 1$, $u_{12} = u_{23} = u_{14} = u_{35} = u_{45} = 1$ with equal weight given to both the tours.

Here we have an example of a slack variable vector U corresponding to a fractional basic feasible solution to Problem 3.4 which need not be an extreme point of $\mathscr{U}(n)$. However, a question that remains is whether the set of all extreme points of $\mathscr{U}(n)$ is the set of all $U's$ corresponding to integer feasible solutions. The answer is NO.

Donald Knuth in [105] states that the Petersen graph is "a remarkable configuration that serves as a counterexample to many optimistic predictions about what might be true for graphs in general." In Example 3.5, the Peterson graph's non-Hamiltonicity is used to our advantage to illustrate our answer.

Example 3.5 Consider Petersen graph $G = (V, E)$ where
$V = \{1, 2, \ldots, 10\}$,
$E = \{(1, 2), (1, 5), (1, 9), (2, 3), (2, 7), (3, 4), (3, 10), (4, 5), (4, 8), (5, 6),$
$(6, 7), (7, 8)\ (8, 9), (9, 10), (6, 10)\}$

Let

$$
d_{ij} = \begin{cases} -1 & \text{if } (i, j) \in E, \\ 0 & \text{otherwise.} \end{cases}
$$

Consider the 10-city STSP on the above graph. It is well known that Petersen graph is non-Hamiltonian i.e. there is no 10-tour available only using the edges of the graph G. Any tour uses 10 edges of the complete graph K_{10}. Thus, an optimal tour for this problem will have an objective value of at least -9 since it has to use an edge not in E.

However, the following fractional solution to the problem has an objective function value of -10.

$x_{134} = x_{135} = x_{356} = x_{147} = x_{178} = x_{348} = x_{478} = x_{139} = x_{189} = x_{389} = x_{3,6,10} = \frac{1}{3}$,

$x_{234} = x_{245} = x_{256} = x_{267} = x_{3910} = \frac{2}{3}$,

$u_{12} = u_{56} = u_{310} = 1$,

$u_{34} = u_{45} = u_{27} = u_{67} = u_{48} = u_{78} = u_{19} = u_{89} = u_{910} = \frac{2}{3}$,

$u_{23} = u_{15} = u_{610} = \frac{1}{3}$; and other u_{ij}'s are zeroes. u_{ij} values are shown along the edges in Fig. 3.1.

As u_{ij}'s add up to 10 and the distance associated with the edges in Petersen graph is -1, we have -10 as the objective function value corresponding to this solution. It is not possible to write this solution as a convex combination of U vectors corresponding to tour solutions, which have an objective function value of at least -9.

Fig. 3.1 Petersen graph with
colour coded values from
Example 3.5

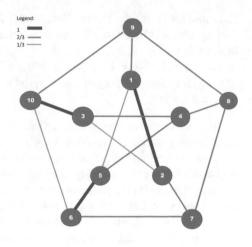

3.4   MI-Relaxation Polytope and Subtour Elimination Polytope

This section presents material from [18]. Let $\mathcal{T}_n$ be the set of all tours in K_n. Then the polytope

$$Q_n = \operatorname{conv}\left\{x_T \in B^{|E_n|} : T \in \mathcal{T}_n\right\}$$

is called the symmetric travelling salesman polytope, where x_T is the c.v. of the tour T, expressed as a set of n edges from E_n. And $\dim(Q_n) = n(n-3)/2$, as shown in [85]. For $n = 3$, $\mathcal{T}_n$ is a singleton. Henceforth, we assume $n \geqslant 4$. Recall the standard formulation for *STSP*.

Definition 3.6 (*Subtour elimination polytope*) Subtour elimination polytope, $SEP(n)$ is the polytope defined by the set of all $u \in R^{|E_n|}$ such that:

$$
\begin{aligned}
& u_e \geqslant 0, \quad \forall e \in E_n \\
& u(\delta(v)) = 2, \quad \forall v \in V_n, \\
& u(\delta(S)) \geqslant 2, \quad \forall \ S \subset V, \ \ 2 \leq |S| \leq n - 2,
\end{aligned}
\tag{3.22}
$$

The dimension of this polytope is $n(n-1)/2$ and the number of constraints is exponential in n. Also, $Q_n \subseteq SEP(n) \ \forall n$.   ♣

Note *Subtours could be removed using the constraints, stated in two different forms:*

$$\text{Either} \quad u(E(S)) \leq |S| - 1, \quad \text{or} \quad u(\delta(S)) \geq 2,$$

where $E(S)$ denotes the set of edges in G with both ends in the node-set S, and $\delta(S)$ denotes the set of edges with one end in S. We are using the cut capacity version. (See Chap. 2, Sect. 2.5 for the definition of cut capacity.)

Theorem 3.2 *Given $n > 3$, and $\mathcal{U}(n)$ as defined by (3.21) and $SEP(n)$ as given by Definition 3.6, we have, $\mathcal{U}(n) \subseteq SEP(n)$.* ♡

Proof The proof is by induction on n. Consider constraints (3.8)–(3.12) of Problem 3.3, other than the non-negativity restrictions (3.11). We introduce the following notation to facilitate the induction proof.

Let u^n_{ij} be the slack variables associated with the constraint corresponding to the edge (i, j) when we have n cities in all. Recall that $\mathcal{U}(n)$ is the set of all U, such that there exists X, such that (X, U) is feasible for Problem 3.3. We have introduced a superscript for U now. Let $U^{(n)}$ be the vector of slack variables $\left(u^n_{ij}\right)$.

We have,

$$\sum_{1 \leq i < j \leq k-1} x_{ijk} = 1, \quad 4 \leq k \leq n, \tag{3.23}$$

$$\sum_{k=4}^{n} x_{ijk} + u^n_{ij} = 1, \quad 1 \leq i < j \leq 3, \tag{3.24}$$

$$-\sum_{r=1}^{i-1} x_{rij} - \sum_{s=i+1}^{j-1} x_{isj} + \sum_{k=j+1}^{n} x_{ijk} + u^n_{ij} = 0, \quad 4 \leq j \leq n-1, 1 \leq i < j, \tag{3.25}$$

$$-\sum_{r=1}^{i-1} x_{rin} - \sum_{s=i+1}^{n-1} x_{isn} + u^n_{i,n} = 0, \quad i = 1, \dots, n-1. \tag{3.26}$$

Now consider the problem with the number of cities equal to $n - 1$, with the first $n - 1$ cities. We have the corresponding equality constraints, after introducing u^{n-1}_{ij}, the slack variables,

$$\sum_{1 \leq i < j \leq k-1} x_{ijk} = 1, \quad 4 \leq k \leq n-1, \tag{3.27}$$

$$\sum_{k=4}^{n-1} x_{ijk} + u^{n-1}_{ij} = 1, \quad 1 \leq i < j \leq 3, \tag{3.28}$$

$$-\sum_{r=1}^{i-1} x_{rij} - \sum_{s=i+1}^{j-1} x_{isj} + \sum_{k=j+1}^{n-1} x_{ijk} + u_{ij}^{n-1} = 0, \quad 4 \leqslant j \leqslant n-2; \ 1 \leqslant i < j,$$

(3.29)

$$-\sum_{r=1}^{i-1} x_{rin} - \sum_{s=i+1}^{n-2} x_{isn} + u_{i,n-1}^{n-1} = 0, \quad i = 1, \ldots, n-2.$$

(3.30)

Comparing these two sets of constraints, we notice that, given a nonnegative solution $\left(X, U^{(n)}\right)$ for the n-city problem, we have, $\left(X/(n-1), U^{(n-1)}\right)$ given below is a non-negative solution to the problem with first $(n-1)$ cities:

$$X/(n-1) = \left(x_{123}, \ldots, x_{n-3,n-2,n-1}\right),$$

(3.31)

$$u_{ij}^{n-1} = u_{ij}^n + x_{ijn}, \ \forall\, 1 \leqslant i < j \leqslant n-2.$$

(3.32)

Basis for induction. We first prove that the result is true for $n = 4$. i.e. $\mathscr{U}(4) \subseteq$ SEP(4). We have u_{ij}^4 as the slack variables. From Eqs. (3.23)–(3.26) we have the following:

$$u_{ij}^4 = 1 - x_{ij4}, \ 1 \leqslant i < j \leqslant 3,$$

(3.33)

$$u_{i4}^4 = \sum_{r=1}^{i-1} x_{ri4} + \sum_{s=i+1}^{3} x_{is4}, \ 1 \leqslant i \leqslant 3.$$

(3.34)

Notice that all u_{ij}^4 are non-negative. Now we show that the degree constraints (3.22) are satisfied for all i.
$i = 1$:

$$u_{12}^4 + u_{13}^4 + u_{14}^4 = 1 - x_{124} + 1 - x_{134} + x_{124} + x_{134} = 2.$$

(3.35)

Similarly checked for $i = 2$ and 3.
$i = 4$:

$$u_{14}^4 + u_{24}^4 + u_{34}^4 = x_{124} + x_{134} + x_{124} + x_{234} + x_{134} + x_{234} = 2.$$

(3.36)

Since the subtour elimination constraints in cut form, for $(V \setminus S)$ are implied by the subtour elimination constraints for S, we verify that the subtour elimination constraints for $|S| = 2$, $S \subseteq V_4$ are satisfied.
Let i_1, i_2, i_3 be a permutation of $(1, 2, 3)$.

$$u^4(\delta(S)) = 2 + 2x_{i_1 i_2 4} \quad \text{for } S = \{i_1, i_2\} \text{ Or } S = \{i_3, 4\}.$$

(3.37)

Thus we have, $u^4(\delta(S)) \geqslant 2$ as $x_{i_1 i_2 4} \geqslant 0$. Hence for $n = 4$, we have the result.

Let us assume $\mathcal{U}(n-1) \subseteq SEP(n-1)$. We shall show that $\mathcal{U}(n) \subseteq SEP(n)$. Since we are going to deal with the value of cut corresponding to subsets of V_n, here on we assume symmetry of the notation of subscripts denoting the edges, i.e, $u_{ij} = u_{ji}$. We show that subtour elimination constraints hold for the required nonempty proper subsets S of V_n i.e. $|S|$ is between 2 and $n/2$. Let

$$\delta^n = \sum_{r \in S, s \in \bar{S}} u^n_{rs} = u^n(\delta(S)), \tag{3.38}$$

be the value of the cut corresponding to a subset S, given $(X, U^{(n)})$ feasible for the n−city problem, with $U^{(n)} \in \mathcal{U}(n)$. We need to show that

$$\delta^n \geq 2 \tag{3.39}$$

Without loss of generality, let $n \in \bar{S}$.
Define

$$P = \{(i, j) \mid x_{ijn} > 0\}. \tag{3.40}$$

Let $S = \{i_1, i_2, \ldots, i_m\}$ and $\bar{S} = \{j_1, j_2, \ldots, j_l, n\}$.
Now consider $U^{(n-1)}$ derived from $U^{(n)}$ and X. We have by the feasibility of $U^{(n-1)}$, $U^{(n-1)} \in \mathcal{U}(n-1)$. And by induction hypothesis $U^{(n-1)} \in SEP(n-1)$. Therefore we have,

$$\delta^{n-1} = \sum_{r=1}^{m} \sum_{s=1}^{l} u^{n-1}_{i_r j_s} \geq 2 \tag{3.41}$$

We need to show that

$$\delta^n = \sum_{r=1}^{m} \sum_{s=1}^{l} u^n_{i_r j_s} + \sum_{r=1}^{m} u^n_{i_r n} \geq 2. \tag{3.42}$$

Take any $i_r \in S$. We have

$$\sum_{s=1}^{l} u^n_{i_r j_s} = \sum_{(i_r, j_s) \in P} u^n_{i_r j_s} + \sum_{(i_r, j_s) \notin P} u^n_{i_r j_s}. \tag{3.43}$$

We have,

$$u^n_{i_r j_s} = \begin{cases} u^{n-1}_{i_r j_s} - x_{i_r j_s n}, & \text{if } (i_r, j_s) \in P, \\ u^{n-1}_{i_r j_s}, & \text{otherwise.} \end{cases}$$

Hence

$$\sum_{s=1}^{l} u_{i_r j_s}^n = \sum_{(i_r, j_s) \in P} u_{i_r j_s}^{n-1} - \sum_{(i_r, j_s) \in P} x_{i_r j_s n} + \sum_{(i_r, j_s) \notin P} u_{i_r j_s}^{n-1}, \tag{3.44}$$

$$u_{i_r n}^n = \sum_{(i_r, i_q) \in P} x_{i_r i_q n} + \sum_{(i_r, j_s) \in P} x_{i_r j_s n}, \tag{3.45}$$

$$\sum_{s=1}^{l} u_{i_r j_s}^n + u_{i_r n}^n = \sum_{(i_r, j_s) \in P} u_{i_r j_s}^{n-1} - \sum_{(i_r, j_s) \in P} x_{i_r j_s n}$$

$$+ \sum_{(i_r, j_s) \notin P} u_{i_r j_s}^{n-1} + \sum_{(i_r, i_q) \in P} x_{i_r i_q n} + \sum_{(i_r, j_s) \in P} x_{i_r j_s n}. \tag{3.46}$$

Therefore,

$$\delta^n = \sum_{r=1}^{m} \left[\sum_{s=1}^{l} u_{i_r j_s}^{n-1} + \sum_{(i_r, i_q) \in P} x_{i_r i_q n} \right] \tag{3.47}$$

$$= \delta^{n-1} + \sum_{r=1}^{m} \sum_{(i_r, i_q) \in P} x_{i_r i_q n} \geqslant 2. \tag{3.48}$$

Hence $\delta^n \geqslant 2 \ \forall n$.

We can check that the degree constraints are satisfied, as follows:

If S is a singleton set, say $S = \{i\}, i \neq n$, then $u^n(\delta(S)) = u^n(\delta(i))$ is still greater than or equal to 2, as the preceding arguments go through for $m = 1$, the cardinality of S. However, notice that for no i the strict inequality can hold, as it will contradict the fact $\sum_{1 \leqslant i < j \leqslant n} u_{ij}^n = n$.

Hence the theorem. $\square$

Formulations of any integer linear programming problem can be compared by using polyhedral information. Suppose two different formulations F_1 and F_2 are stated in the same space of variables $x \in R^p$ and the problem is to find an optimal integer solution that minimises the objective function. Let $P(F_1)$ and $P(F_2)$ be the polyhedra associated with these formulations. If $P(F_1) \subset P(F_2)$, then F_1 is a better formulation than F_2 since the lower bound obtained by solving the LP relaxation of F_1 is as good as or better than the one obtained by solving the LP relaxation of F_2. Consider the case in which F_3 is a formulation having the polyhedron

$$P(F_3) = \{(x, y) \in R^p \times R^q \mid Ax + By \leq b\},$$

where A, B and b have same number of rows. To compare the formulation F_3 with the formulation F_1, we can use the projection of $P(F_3)$ into the subspace of x variables given by

$$TP(F_3) = \{x \in R^p : \exists \, y \in R^q \ni (x, y) \in P(F_3)\}.$$

That is, if $TP(F_3) \subset P(F_1)$, then F_3 is a better formulation.

There is an analogy of catching the lion by fencing around it and optimisation over polytopes. Ellipsoidal algorithms and branch-and-cut schemes start with an outer polytope (fence) that contains the polytope of interest and successively cut off regions of the outer polytope (tighten the fence) until an optimal solution is found (the lion is caught). Thus, an important fact that follows from Theorem 3.2 is MI-relaxation provides as good a 'fence' as $SEP(n)$.

 Note $Q_n \subset \mathcal{U}(n) \subseteq SEP(n)$, *and hence MI-formulation is as tight as $SEP(n)$ when our interest is solving STSP. There are relaxations of other formulations of STSP which are as tight of $SEP(n)$ (see [17] and references cited therein).*

3.4.1 What Next?

This chapter introduced a multistage decision problem related to *STSP* using a dynamic programming approach, this is different from the one given by Bellman or Held and Karp we saw in Chap. 1. Next, we visualised the insertion heuristic-inspired problem as a combinatorial optimisation problem. To better understand the structure of this *COP* we identified equalities and inequalities met by the objects of interest, namely, the feasible decision vectors. A 0/1 programming problem called *MI*-formulation is given and showed this correctly solves the *STSP*.

In addition, we have demonstrated that the set of slack variable vectors corresponding to feasible solutions to the *MI*-relaxation is contained in the subtour elimination polytope defined by exponentially many constraints.

Interesting, but, what is the connection between pedigrees and our objects of interest? You may ask.

The feasible decision vectors are pedigrees.

The Fig. 3.2 illustrates this. Example 3.6 considers an instance with $n = 7$, to bring out the $1 - 1$ correspondence between the pedigree, the feasible decision vector and the 7-tour. A formal proof will be given in Chap. 4, where we explore pedigree polytopes further.

Example 3.6 Consider $n = 7$. Recall the definition of pedigree from Sect. 1.3.2. Our initial triangle is $\{1, 2, 3\}$. The decision to insert 4 in $(1, 3)$ corresponds to the selection of triangle $\{1, 3, 4\} \in \Delta^4$. The feasible decision vector corresponding to the edges selected for inserting 4, 5, 6 and 7 are given by $W = ((1, 3), (3, 4), (2, 3), (1, 2))$ in that order. Notice that the resulting 7-tour is the boundary of the triangles in the pedigree. Also, since $\{1, 2, 3\}$ is in all pedigrees with weight 1, we do not consider it in the *MI*-formulation, as it means to increase the number of variables by 1 and have the additional constraint $x_{123} = 1$.

Fig. 3.2 Pedigree, Feasible
Decision vector, Tour. The
7-tour is shown in brown.
The edges chosen for
insertion are coloured to
match the common facets of
the triangles

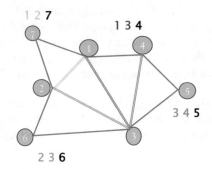

Edges for insertion: W = ((1, 3), (3, 4), (2, 3), (1, 2))

Feasible Decision vector:

X = (010, 000001, 0010000000, 100000000000000)

 Note *In Example 3.6 observe the* $1 - 1$ *correspondence between (1) pedigree as defined in Sect. 1.5, (2) feasible decision vector which is an integer solution to MI-formulation, and (3) the 7-tour. This fact is in general true.*

Thus, we can equivalently define a pedigree as follows:

> **Definition 3.7** *(Pedigree)* $W = (e_4, \ldots, e_n) \in E_3 \times \cdots \times E_{n-1}$ is called a *pedigree* if and only if there exists a $T \in \mathcal{T}_n$ such that T is obtained from the $3 - tour$ by the sequence of insertions, viz.,
>
> $$ 3 - tour \; \overrightarrow{e_4, 4} \;\; T^4 \ldots T^{n-1} \overrightarrow{e_n, n} \;\; T. $$
>
> ♣

The pedigree W is referred to as the pedigree of T.

The pedigree of T can be obtained by shrinking T sequentially to the $3 - tour$ and noting the edge created at each stage. We then write the edges obtained in the reverse order of their occurrence. Let the set of all pedigrees for a given n be denoted by $\mathcal{P}_n$. We can associate a $X = (x_4, \ldots, x_n) \in B^{T_n}$, the characteristic vector of the pedigree W, where $(W)_k = e_k$, the $(k - 3)^{rd}$ component of W, $4 \leq k \leq n$ and x_k is the indicator of e_k. Notice that feasible decision vectors and integer solutions to MI-formulation are the same as indicators of pedigrees.

Definition 3.8 (*Pedigree Polytope*) Let $P_n = \{X \in B^{\tau_n} : X$ is the characteristic vector of the pedigree $W \in \mathcal{P}_n\}$.

Consider the convex hull of P_n. We call this the *pedigree polytope* and denote it by $conv(P_n)$. ♣

Thus, there is a one-to-one correspondence between $T \in \mathcal{T}_n$ and $X \in P_n$.

In Chap. 4 we have one more definition of pedigree. The advantage of the different definitions of pedigree is seen while proving results on the structure of the pedigree polytope, and in proving results connecting *STSP* polytope and pedigree polytope.

Chapter 4
Structure of the Pedigree Polytope

4.1 Introduction

In Chap. 3 I promised that I will be exploring further into the pedigree polytope. Recall the definition of a pedigree, and pedigree polytope, the convex hull of the set of pedigrees. This is a $\mathcal{V}$-presentation of the pedigree polytope. As we saw in Chap. 2, in the section on polytopes, as well as in Chap. 1 while discussing *STSP* polytope, $\mathcal{V}$-presentation is not much of use in reducing non-determinism, unless we use our understanding of the characteristics of the underlying polytope (recall the success of branch-and-cut strategy reported by William Cook for *STSP* [51]). With the view to characterising the pedigree polytope, we study the *MI*-relaxation problem. As we have already observed in Chap. 3, any pedigree satisfies all the constraints of the *MI*-relaxation problem, in addition, to be a 0/1 vector. So it makes sense to further explore *MI*-relaxation polytope.

In this chapter, I characterise the pedigree polytope by proving a necessary and sufficient condition for membership in it. The adjacency structure of the vertices of a polytope is another useful area of study which reveals the geometry of a polytope. We show in this chapter, a complete characterisation of the adjacency (nonadjacency) structure of the pedigree polytope. Also, I give a low order strongly polynomial algorithm to check the nonadjacency of pedigrees.

4.2 MI-Relaxation Polytope

Consider the *MI*-formulation discussed in Chap. 3. We now elaborate on the corresponding linear relaxation problem in this section.

Problem 4.1 (*MI-Relaxation*)

$$minimise \sum_{k=4}^{n} \sum_{1 \leq i < j \leq k-1} c_{ijk} x_{ijk}$$

subject to

$$\sum_{1 \leq i < j \leq k-1} x_{ijk} = 1, \quad 4 \leq k \leq n \tag{4.1}$$

$$\sum_{k=4}^{n} x_{ijk} \leq 1, \quad 1 \leq i < j \leq 3 \tag{4.2}$$

$$-\sum_{r=1}^{i-1} x_{rij} - \sum_{s=i+1}^{j-1} x_{isj} + \sum_{k=j+1}^{n} x_{ijk} \leq 0, \quad 4 \leq j \leq n-1; 1 \leq i < j \tag{4.3}$$

$$-\sum_{r=1}^{i-1} x_{rin} - \sum_{s=i+1}^{n-1} x_{isn} \leq 0, \quad i = 1, \ldots, n-1. \tag{4.4}$$

$$x_{ijk} \geq 0, \quad 4 \leq k \leq n; \tag{4.5}$$
$$1 \leq i < j \leq k-1$$

We refer to Problem 4.1 as *MI*-relaxation, and refer to it as problem *MI(n)*. Observe that for each (i, j) we have an inequality constraint and there are p_n of them. And constraints (4.4) are redundant. Let the slack variable corresponding to the inequality constraint for (i, j) be given by $U^{(n)} = (u_{ij}^n)$.

 Note *We have already exploited the recursive structure of the MI-relaxation while proving Theorem 3.2. We further prove results based on that, and formalise the notation here, by introducing n in the superscript of U, showing the problem's dependency on n.*

Given $n \geq 4$, X, an integer solution to Problem 4.1, the slack variable vector, $U^{(n)} \in B^{p_n}$ gives the edge-tour incident vector of the corresponding n-tour, as proved earlier, by Lemma 3.1. Next, we make the recursive structure of *MI*-relaxation explicit.

Definition 4.1 In general, let $E_{[n]}$ denote the matrix corresponding to Eq. (4.1); let $A_{[n]}$ denote the matrix corresponding to the inequalities (4.2), (4.3) and (4.4). Let $\mathbf{1}_r$ denote the row vector of r $1's$. Let I_r denote the identity matrix of size $r \times r$.

$$E_{[n]} = \begin{pmatrix} \mathbf{1}_{p_3} & \cdots & 0 & 0 \\ \vdots & \vdots & \vdots & \vdots \\ 0 & \cdots & \mathbf{1}_{p_{n-2}} & 0 \\ 0 & 0 & \cdots & \mathbf{1}_{p_{n-1}} \end{pmatrix} = \begin{pmatrix} E_{[n-1]} & 0 \\ 0 & \mathbf{1}_{p_{n-1}} \end{pmatrix}.$$

To derive a recursive expression for $A_{[n]}$ we first define

$$A^{(n)} = \begin{pmatrix} I_{p_{n-1}} \\ -M_{n-1} \end{pmatrix}$$

where M_i is the $i \times p_i$ *node-edge incidence* matrix of K_i, the complete graph on V_i, and the edges (i, j) are ordered in the ascending order of the edge labels, l_{ij}.

Then

$$A_{[n]} = \begin{pmatrix} A^{(4)} & | & A^{(5)} & | & & | & A^{(n)} \\ & | & & | & \ddots & | & \\ \mathbf{0} & | & \mathbf{0} & | & & | & \end{pmatrix} = \begin{pmatrix} A_{[n-1]} & | & A^{(n)} \\ \mathbf{0} & | & \end{pmatrix}.$$

Observe that $A^{(n)}$ is the sub matrix of $A_{[n]}$ corresponding to $\mathbf{x}_n$. The number of rows of $0's$ is decreasing from left to right.   ♣

We can state Problem 4.1 in matrix notation for a given $l \geq 4$ as Problem 4.2. Let $U^{(3)} = \begin{pmatrix} \mathbf{1}_3 \\ \mathbf{0} \end{pmatrix}$, where the $\mathbf{0}$ is a column vector of $p_l - 3$ zeros.

Problem 4.2 (*Problem MI(l)*) Find $X \in R^{\eta}, U^{(l)} \in R^{p_l}$ solution to

$$minimise \; CX$$

subject to

$$E_{[l]}X = \mathbf{1}_{l-3}$$
$$A_{[l]}X + U^{(l)} = U^{(3)}$$
$$X \geq \mathbf{0}$$
$$U^{(l)} \geq \mathbf{0}$$

Expanding the l.h.s of the equations in Problem 4.2 for l using the recursive structure of $E_{[l]}$ and $A_{[l]}$ and replacing $X = (X/l - 1, \mathbf{x}_l)$ we can observe,

$$E_{[l-1]}X/l - 1 = \mathbf{1}_{l-4} \tag{4.6}$$

$$\mathbf{1}_{p_{l-1}}\mathbf{x}_l = 1 \tag{4.7}$$

$$\begin{pmatrix} A_{[l-1]} \\ \mathbf{0} \end{pmatrix} X/l - 1 + A^{(l)}\mathbf{x}_l + U^{(l)} = U^{(3)} \tag{4.8}$$

Or from the definition of $U^{(l-1)}$ from Problem $MI(l-1)$, we have

$$U^{(l-1)} - A^{(l)}\mathbf{x}_l = U^{(l)}, \ \forall\, l, 4 \le l \le n. \tag{4.9}$$

Property 4.1 (*Stem Feasibility*) Let $n \ge 5$. If $X \in R^{\tau_n}$ is feasible for Problem $MI(n)$ implies X/l is feasible for Problems $MI(l)$ for $l \in \{4, \ldots, n-1\}$.

Proof Suppose the property is not true. Look at the l for which X/l is not feasible for Problem $MI(l)$. Then there exists a $U_{ij}^{(l)} < 0$ for some ij. This implies $U_{ij}^{(n)} < 0$ as well since $\sum_{k=l+1}^{n} x_{ijk} \ge 0$. So X is not feasible for Problem $MI(n)$ leading to a contradiction. So there is no such l, or the property is true.

Given n, $Y \in R^{\tau_n}$, using $U^{(l)}$, the slack variable vectors in Problem $MI(l)$, for $l, \in \{4, \ldots, n\}$ (with trailing zeros where needed), we can reformulate the constraints of the MI-relaxation problem in matrix notation as follows :

$$E_{[n]}Y = \mathbf{1}_{n-3} \tag{4.10}$$

$$U^{(3)} = \begin{pmatrix} \mathbf{1}_3 \\ \mathbf{0} \end{pmatrix} \tag{4.11}$$

$$U^{(l-1)} - A^{(l)}\mathbf{y}_l = U^{(l)}, \ for \ all \ l, \ 4 \le l \le n. \tag{4.12}$$

$$U \ge 0, \tag{4.13}$$

$$Y \ge 0. \tag{4.14}$$

We explain the notations introduced using the following example.

Example 4.1 Consider $n = 5$, then $p_4 = 6$, $p_5 = 10$, and $\tau_n = 9$. Let Y be given by,

$$Y' = (1/2, 0, 1/2; \ 1/2, 0, 0, 0, 0, 1/2) \in R^{\tau_5}.$$

where Y' denotes the transpose of Y. Y satisfies the equality restriction (4.1) of the MI-relaxation. Also, the corresponding slack variable vector obtained from the rest of the inequalities (4.2), (4.3) and (4.4) is nonnegative. We have,

$$U^{(5)} = \begin{pmatrix} \mathbf{1}_3 \\ \mathbf{0} \end{pmatrix} - A_{[5]}Y = (0, 1, 1/2, 1/2, 1, 0, 1/2, 1/2, 1/2, 1/2).$$

(Matrices $E_{[5]}$ and $A_{[5]}$ are shown in Fig. 4.1.)

Fig. 4.1 Matrices $E_{[5]}$ and $A_{[5]}$

$$E_{[5]} = \begin{pmatrix} 1 & 1 & 1 & 0 & 0 & 0 & 0 & 0 & 0 \\ 0 & 0 & 0 & 1 & 1 & 1 & 1 & 1 & 1 \end{pmatrix},$$

$$A_{[5]} = \begin{pmatrix} 1 & 0 & 0 & 1 & 0 & 0 & 0 & 0 & 0 \\ 0 & 1 & 0 & 0 & 1 & 0 & 0 & 0 & 0 \\ 0 & 0 & 1 & 0 & 0 & 1 & 0 & 0 & 0 \\ -1 & -1 & 0 & 0 & 0 & 0 & 1 & 0 & 0 \\ -1 & 0 & -1 & 0 & 0 & 0 & 0 & 1 & 0 \\ 0 & -1 & -1 & 0 & 0 & 0 & 0 & 0 & 1 \\ 0 & 0 & 0 & -1 & -1 & 0 & -1 & 0 & 0 \\ 0 & 0 & 0 & -1 & 0 & -1 & 0 & -1 & 0 \\ 0 & 0 & 0 & 0 & -1 & -1 & 0 & 0 & -1 \\ 0 & 0 & 0 & 0 & 0 & 0 & -1 & -1 & -1 \end{pmatrix}.$$

So $(Y, U^{(5)})$ is a feasible solution to the *MI*-relaxation. Here, $U^{(4)}$ is given by $(1/2, 1, 1/2, 1/2, 1, 1/2)'$.

Lemma 4.1 *The MI-relaxation problem, is in the combinatorial LP class.* ♡

Proof Given n, the *MI-relaxation* problem, has the matrix A with entries from $\{+1, -1 \text{ or } 0\}$, and the dimension of the input is bounded above by $p_n \times \tau_n$. So theoretically, we are assured of a strongly polynomial algorithm to solve any instance of the *MI*-relaxation problem [159].

4.3 A Polytope That Contains Pedigree Polytope

Let $X = (\mathbf{x}_4, \ldots, \mathbf{x}_n) \in P_n$ correspond to the pedigree $W = (e_4, \ldots, e_n)$. I state the following results without proof:

Lemma 4.2 $X \in P_n$ *implies* $X \geq 0$ *and*

$$x_k(E_{k-1}) = 1, k \in V_n \setminus V_3. \tag{4.15}$$

♡

Lemma 4.3 $X \in P_n$ implies

$$\sum_{k=4}^{n} x_k(e) \le 1,\ e \in E_3. \tag{4.16}$$

$\heartsuit$

Lemma 4.4 $X \in P_n$ implies

$$-x_j(\delta(i) \cap E_{j-1}) + \sum_{k=j+1}^{n} x_k(e) \le 0,\ e = (i, j) \in E_{n-1} \setminus E_3. \tag{4.17}$$

$\heartsuit$

Notice that the equations and inequalities considered in the above lemmas are the same as that of the *MI*-relaxation, except that we have a three subscripted notation and some redundant constraints in the *MI*-relaxation.

Definition 4.2 (P_{MI} (n)-polytop) Consider $X \in R^{T_n}$ satisfying the non negativity restrictions, $X \ge 0$, the Eq. (4.15) and the inequalities (4.16) and (4.17).

The set of all such X is a polytope, as we have defined it using linear equalities and inequalities. We call this polytope $P_{MI}(n)$.   ♣

As every pedigree satisfies the equalities and inequalities that define $P_{MI}(n)$, we can now conclude that $conv(P_n) \subset P_{MI}(n)$.

Checking whether $X \in P_{MI}(n)$ can be done by checking whether X is a feasible solution to *MI*-relaxation. And it can be done recursively as follows:

Lemma 4.5 *If $Y \in P_{MI}(n)$, then there exists U satisfying Eqs. 4.11 through 4.13. Conversely, if U and Y satisfy Eqs. 4.10 through 4.14 then $Y \in P_{MI}(n)$.*

$\heartsuit$

Proof Follows from Property 4.1.

4.3.1   Alternative Definition of a Pedigree

I defined Pedigree in Chap. 1 using 2-simplex or triangles, and again in Chap. 3 using n-tours.

In this subsection, I define a pedigree without reference to a Hamiltonian cycle, which makes explicit an interesting property of pedigrees.

Definition 4.3   (*Generators of an edge*) Given $e_\beta = (i, j) \in E_k$, we say $G(e_\beta)$ is the set of generators of e_β in case

$$G(e_\beta) = \begin{cases} E_3 \setminus \{e_\beta\} & \text{if } e_\beta \in E_3 \\ \delta_i \cap E_{j-1} & \text{otherwise.} \end{cases}$$

♣

Remark   Since an edge $e = (i, j)$, $j > 3$ is generated by inserting j in any e' in the set $G(e)$, the name *generator* is used to denote any such edge.

Caution: Though we define an edge as a generator of (i, j) it is implicit that we are referring to the decision of inserting $j > 3$, in an edge e' to obtain (i, j). Hence the triangle $\{i, j, k\}$ has the same info as the insertion of k in (i, j), and an ordered set of edges can be used instead, to represent a pedigree.

An equivalent definition of a pedigree is possible as stated in Lemma 4.6 proved in [14].

Lemma 4.6   *Given n, consider $W = (e_4, \ldots, e_n)$, where $e_k = (i_k, j_k)$ for $1 \le i_k < j_k \le k - 1, 4 \le k \le n$. W corresponds to a pedigree in $\mathcal{P}_n$ if and only if*

(a) $e_k, 4 \le k \le n$, are all distinct,
(b) $e_k \in E_{k-1}, 4 \le k \le n$, and
(c) for every $k, 5 \le k \le n$, there exists a $e' \in G(e_k)$ such that, $e_q = e'$, where $q = \max\{4, j_k\}$.

♡

This lemma allows us to define a pedigree without explicitly considering the corresponding Hamiltonian cycle.

Definition 4.4   (*Extension of a Pedigree*) Let $y(e)$ be the indicator of $e \in E_k$. Given a pedigree, $W = (e_4, \ldots, e_k)$ (with the characteristic vector, $X \in P_k$) and an edge $e \in E_k$, we call $(W, e) = (e_4, \ldots, e_k, e)$ an *extension* of W in case $(X, y(e)) \in P_{k+1}$.

♣

Using Lemma 4.6, observe that given W a pedigree in $\mathcal{P}_k$ and an edge $e = (i, j) \in E_k$, (W, e) is a pedigree in $\mathcal{P}_{k+1}$ if and only if 1] $e_l \neq e$, $4 \leq l \leq k$ and 2] there exists a $q = max(4, j)$ such that e_q is a generator of $e = (i, j)$.

This definition comes in handy in proving certain results later in the book.

4.4   Characterisation Theorems

In this section, given a $X \in P_{MI}(n)$ we wish to know whether X is indeed in $conv(P_n)$. Some of these results presented here find use in Sect. 4.5. Let $|P_k|$ denote the cardinality of P_k. Assume that the pedigrees in $\mathcal{P}_k$ are numbered (say, according to the lexicographical ordering of the edge labels of the edges appearing in a pedigree).

Definition 4.5 Given $X = (\mathbf{x}_4, \ldots, \mathbf{x}_n) \in P_{MI}(n)$ we denote by $X/k = (\mathbf{x}_4, \ldots, \mathbf{x}_k)$, the restriction of X, for $4 \leq k \leq n$.

Given $X \in P_{MI}(n)$ and $X/k \in conv(P_k)$, consider $\lambda \in R_+^{|P_k|}$ that can be used as a weight to express X/k as a convex combination of $X^r \in P_k$. Let $I(\lambda)$ denote the index set of positive coordinates of λ. Let $\Lambda_k(X)$ denote the set of all possible *weight vectors*, for a given X and k, that is,

$$\Lambda_k(X) = \left\{ \lambda \in R_+^{|P_k|} \mid \sum_{r \in I(\lambda), X^r \in P_k} \lambda_r X^r = X/k, \sum_{r \in I} \lambda_r = 1 \right\}.$$

♣

Definition 4.6 Consider a $X \in P_{MI}(n)$ such that $X/k \in conv(P_k)$. We denote the $k - tour$ corresponding to a pedigree X^α by T^α. Given a weight vector $\lambda \in \Lambda_k(X)$, we define a *FAT* problem with the following data:

$$\begin{aligned}
O -- \quad & Origins] : \alpha, \alpha \in I(\lambda) \\
a -- \quad & Supply] : a_\alpha = \lambda_\alpha \\
D -- \quad Destinations] : \; & \beta, e_\beta \in E_k, x_{k+1}(e_\beta) > 0 \\
b -- \quad & Demand] : b_\beta = x_{k+1}(e_\beta) \\
\mathcal{A} -- \quad & Arcs] : \{(\alpha, \beta) \in O \times D | e_\beta \in T^\alpha\}
\end{aligned}$$

We designate this problem as $FAT_k(\lambda)$. Notice that arcs (α, β) not satisfying $e_\beta \in T^\alpha$ are the forbidden arcs. We also say FAT_k is feasible if the problem $FAT_k(\lambda)$ is feasible for some $\lambda \in \Lambda_k(X)$.

♣

Fig. 4.2 FAT_4 Problem for Example 4.2

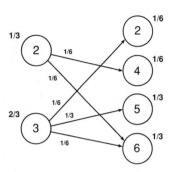

Equivalently, the arcs in $\mathcal{A}$ can be interpreted as follows: If W^α is the pedigree corresponding to $X^\alpha \in P_k$ for an $\alpha \in I(\lambda)$ then the arcs $(\alpha, \beta) \in \mathcal{A}$ are such that the (W^α, e_β) is an extension of W^α, that is, $(W^\alpha, e_\beta) \in \mathscr{P}_{k+1}$.

Example 4.2 Consider $X = (0, \frac{1}{3}, \frac{2}{3}, 0, \frac{1}{6}, 0, \frac{1}{6}, \frac{1}{3}, \frac{1}{3})$. We wish to check whether X is in $conv(P_5)$. It is easy to check that X indeed satisfies the constraints of $P_{MI}(5)$. Also $X/4 = (0, \frac{1}{3}, \frac{2}{3})$ is obviously in $conv(P_4)$. And $\Lambda_4(X) = \{(0, \frac{1}{3}, \frac{2}{3})\}$. Assume that the pedigrees in $\mathcal{P}_4$ are numbered such that, $X^1 = (1, 0, 0)$, $X^2 = (0, 1, 0)$ and $X^3 = (0, 0, 1)$ and the edges in E_4 are numbered according to their edge labels. Then $I(\lambda) = \{2, 3\}$. Here $k = 4$ and the $FAT_4(\lambda)$ is given by a problem with origins, $O = \{2, 3\}$ with supply $a_2 = \frac{1}{3}, a_3 = \frac{2}{3}$ and destinations, $D = \{2, 4, 5, 6\}$ with demand $b_2 = b_4 = \frac{1}{6}, b_5 = b_6 = \frac{1}{3}$. Corresponding to origin 2 we have the pedigree $W^2 = ((1, 3))$. And the edge corresponding to destination 2 is $e_2 = (1, 3)$. As (W^2, e_2) is not an extension of W^2, we do not have an arc from origin 2 to destination 2. Similarly, (W^3, e_4), (W^2, e_5) are not extensions of W^3 and W^2 respectively, so we do not have arcs from origin 3 to destination 4 and origin 2 to destination 5. We have the set of arcs given by,

$$\mathcal{A} = \{(2, 4), (2, 6), (3, 2), (3, 5) \text{ and } (3, 6)\}.$$

Notice that f given by $f_{24} = f_{26} = f_{32} = f_{36} = \frac{1}{6}$, $f_{35} = \frac{1}{3}$ is feasible to $FAT_4(\lambda)$. (See Fig. 4.2). This f gives a weight vector to express X as a convex combination of the vectors in P_5, which are the extensions corresponding to arcs with a positive flow. This role of f is in general true and we state this as Theorem 4.1.

It is easy to check that f is the unique feasible flow in this example, so no other weight vector exists to certify X in $conv(P_5)$. Thus we have expressed X as a convex combination of the incidence vectors of the pedigrees $W^7 = ((1, 3), (1, 4))$, $W^8 = ((1, 3), (3, 4))$, $W^{10} = ((2, 3), (1, 3))$, $W^{12} = ((2, 3), (3, 4))$ (each of them receive a weight of $\frac{1}{6}$) and $W^{11} = ((2, 3), (2, 4))$ (which receives a weight of $\frac{1}{3}$).

Theorem 4.1 *Let $k \in V_{n-1} \setminus V_3$. Suppose $\lambda \in \Lambda_k(X)$ is such that $FAT_k(\lambda)$ is feasible. Consider any feasible flow f for the problem. Let $W^{\alpha\beta}$ denote the extension (W^α, e_β), a pedigree in $\mathcal{P}_{k+1}$, corresponding to the arc (α, β). Let $\mathcal{W}_f$ be the set of such pedigrees, $W^{\alpha\beta}$ with positive flow $f_{\alpha\beta}$. Then f provides a weight vector to express $X/k + 1$ as a convex combination of pedigrees in $\mathcal{W}_f$.* ♡

Next, we observe that $conv(P_n)$ can be characterized using a sequence of flow feasibility problems as stated in the following theorems:

Theorem 4.2 *If $X \in conv(P_n)$ then FAT_k is feasible $\forall k \in V_{n-1} \setminus V_3$.* ♡

Proof Since $X \in conv(P_n)$ we have

$$x_l(e) = \sum_{s \in I(\lambda)} \lambda^s x_l^s(e), e \in E_{l-1}, l \in V_{n-1} \setminus V_3$$

where $I(\lambda)$ is the index set for some $\lambda \in \Lambda_k(X)$.

We say a pedigree in $X \in P_n$ is a descendant of a pedigree in $Y \in P_k$ in case $X/k = Y$. To show that X is such that FAT_k are all feasible, we proceed as follows: First partition $I(\lambda)$ according to the pedigree, $X^\alpha \in P_k$,

$$S_O^\alpha = \{s | s \in I(\lambda) \text{ and } X^s \text{ is a descendant of } X^\alpha \in P_k\}. \quad (4.18)$$

Secondly, partition $I(\lambda)$ according to the edge $e_\beta \in E_k$,

$$S_D^\beta = \{s | s \in I(\lambda) \text{ and } X^s \text{ is such that } x_{k+1}^s(e_\beta) = 1\}. \quad (4.19)$$

O and D in the suffixes refer to origins and destinations in the *FAT* problem. Let $a_\alpha = \sum_{s \in S_O^\alpha} \lambda^s$; $b_\beta = \sum_{s \in S_D^\beta} \lambda^s = x_{k+1}(e_\beta)$ and let the set of arcs be given by

$$\mathcal{A} = \{(\alpha, \beta) | S_O^\alpha \cap S_D^\beta \neq \emptyset\}.$$

Now for $k = 4$ we have $(X/4) \in P_4$, as a_α are positive for $\alpha \in O$ and add up to 1. Applying Lemma 2.1, we have a feasible f for this problem given by $f_{\alpha\beta} = \sum_{s \in S_O^\alpha \cap S_D^\beta} \lambda^s$. Thus FAT_4 is feasible. And so $X/5$ is in $conv(P_5)$. Also, notice that the origins corresponding to $FAT_5(\lambda_5)$ are precisely the pedigrees in P_5 with $f_{\alpha\beta}$ positive, here λ_5 is given by the feasible flow for FAT_4 given by Lemma 2.1.

In general, let λ_k be defined as $\lambda_k(X^\alpha) = a_\alpha$, $X^\alpha \in P_k$ and $S_O^\alpha \neq \emptyset$. In the $FAT_k(\lambda_k)$ problem, we have an origin for each pedigree in P_k which receives a positive weight according to λ_k. We have a *FAT* problem, corresponding to k and λ_k. The feasibility of *FAT$_k$* for every $k \in V_{n-1} \setminus V_3$ then follows from an application of Lemma 2.1. Hence the theorem. $\qquad\square$

Theorem 4.3 *Let $k \in V_{n-1} \setminus V_3$. If $\lambda \in \Lambda_k(X)$ is such that FAT$_k(\lambda)$ is feasible, then $X/(k+1) \in conv(P_{k+1})$.* $\qquad\heartsuit$

Proof Consider X^α, $\alpha \in I(\lambda)$ and an edge $e_\beta \in E_k$ such that (α, β) is not forbidden. So $e_\beta \in T^\alpha \in \mathcal{H}_k$. So e_β is one of the edges available for insertion of $k+1$. As noticed earlier, every (α, β) not forbidden corresponds to a pedigree $X^{\alpha\beta} \in P_{k+1}$ as defined below:

$$X^{\alpha\beta} = (X^\alpha, y^\beta) \text{ where } y^\beta(e) = \begin{cases} 1 & \text{if } e = e_\beta \\ 0 & \text{otherwise.} \end{cases}$$

(y^β is the indicator of e_β.) Therefore, from the feasibility of *FAT$_k(\lambda)$* we have a flow with $f_{\alpha\beta} \geq 0$, and

$$\sum_{\beta\ni(\alpha,\beta)\in\mathcal{A}} f_{\alpha\beta} = \lambda_\alpha, \ \alpha \in I(\lambda) \tag{4.20}$$

$$\sum_{\alpha\ni(\alpha,\beta)\in\mathcal{A}} f_{\alpha\beta} = x_{k+1}(e_\beta), \ e_\beta \in E_k, \ x_{k+1}(e_\beta) > 0. \tag{4.21}$$

We shall show that
$$\sum_\alpha \sum_\beta X^{\alpha\beta} f_{\alpha\beta} = X/k + 1. \tag{4.22}$$

Substituting $X^{\alpha\beta} = (X^\alpha, y^\beta)$ in the above equation and simplifying we get

$$\sum_\alpha \sum_\beta X^{\alpha\beta} f_{\alpha\beta} = \left(\sum_\alpha X^\alpha \sum_\beta f_{\alpha\beta}, \sum_\beta y^\beta \sum_\alpha f_{\alpha\beta} \right) \tag{4.23}$$

$$= \left(\sum_\alpha \lambda_\alpha X^\alpha, \sum_\beta x_{k+1}(e_\beta)y^\beta \right) \tag{4.24}$$

$$= (X/k, \mathbf{x}_{k+1})$$
$$= X/k + 1.$$

Hence the result. $\qquad\square$

Remark [1] In general we do not have to explicitly give the set $\Lambda_k(X)$, or work with pedigree numbers as done in Example 4.2. The set $\Lambda_k(X)$ or pedigree numbers are used in proofs only. [2] The constructions of S_O^α and S_D^β are useful and will be used later in defining a *FAT* problem in Chap. 5 (Definition 5.6).

Theorem 4.4 (Necessary and Sufficient Condition) *Thus, for a given $X \in P_{MI}(n)$ the condition*

$$\forall k \in V_{n-1} \setminus V_3, \; \exists \, a \, \lambda \in \Lambda_k(X) \text{ such that } FAT_k(\lambda) \text{ is feasible}$$

is both necessary and sufficient for X to be in $conv(P_n)$.   ♡

In Theorem 4.3 we have a procedure to check whether a given $X \in P_{MI}(n)$, is in the pedigree polytope, $conv(P_n)$. Since feasibility of a $FAT_k(\lambda)$ problem for a weight vector λ implies $X/(k+1)$ is in $conv(P_{k+1})$, we can sequentially solve $FAT_k(\lambda_k)$ for each $k = 4, \ldots, n-1$ and if $FAT_k(\lambda_k)$ is feasible we set $k = k+1$ and while $k < n$ we repeat; at any stage if the problem is infeasible we stop. So if we have reached $k = n$ we have proof that $X \in conv(P_n)$. However if for a $\lambda \in \Lambda_k(X)$ the problem is infeasible we cannot conclude that $X \notin conv(P_n)$. Example 4.3 illustrates this.

Example 4.3 Consider X given by

$$\mathbf{x}_4 = (1/2, 1/2, 0);$$
$$\mathbf{x}_5 = (0, 0, 1/2, 1/2, 0, 0);$$
$$\mathbf{x}_6 = (0, 0, 0, 1/2, 1/2, 0, 0, 0, 0, 0).$$

$FAT_4(\lambda)$ for the unique $\lambda = \mathbf{x}_4$ is feasible. f given by $f((1, 2), (2, 3)) = f((1, 3), (1, 4)) = \frac{1}{2}$ with the flow along other arcs zero is a feasible flow for $FAT_4(\lambda)$.

Now the problem $FAT_5(\lambda)$ corresponding to the λ given by f is infeasible as the maximum flow in the corresponding network is only $\frac{1}{2}$.

We are not able to conclude whether $X \in conv(P_6)$ or not. But we can check that $X = \frac{1}{2}(X^1 + X^2)$ where X^1 is given by $x_4^1((1, 2)) = x_5^1((1, 4)) = x_6^1((2, 4)) = 1$ and X^2 is given by $x_4^2((1, 3)) = x_5^2((2, 3)) = x_6^2((1, 4)) = 1$.

However if we have chosen the alternative f^*, feasible solution for $FAT_4(\lambda)$, given by $f^*((1, 2), (1, 4)) = f^*((1, 3), (2, 3)) = \frac{1}{2}$ with flow along other arcs zero, we have the problem $FAT_5(\lambda^*)$ corresponding to f^*. And this problem is feasible and so we conclude $X \in conv(P_6)$. How complex it is, in general, to check membership in $conv(P_n)$ is the main concern of this book and Chaps. 5 and 6 discuss this issue.

To explore further the structure of the pedigree polytope, I shall discuss the adjacency structure in the 1-skeleton of the polytope.

4.5 Nonadjacency in Pedigree Polytopes

Papadimitriou [138] has shown that the nonadjacency testing on the Travelling Sales-man Polytope is **NP-complete**. The asymmetric version of the problem was con-sidered by Heller in 1955 [94]. Katta. G. Murty [126] gave a purported necessary and sufficient condition for two tours to be nonadjacent in the convex hull of tours. Theorem 2 in [126], states that $T^{[1]}$ and $T^{[2]}$ are nonadjacent if and only if there exists a tour $T^{[3]}$ different from both $T^{[i]}$, $i = 1, 2$ such that

$$T^{[3]} \subset \cup_{i=1,2} T^{[i]} \text{ and } \cap_{i=1,2} T^{[i]} \subset T^{[3]}.$$

M. Rammohan Rao [145] gives a counter example to show that this condition, though a necessary one, is not sufficient.

What can we say about the corresponding nonadjacency testing problem for pedi-grees? We state this problem and study its complexity in this section, which is very promising compared to that of Q_n. An interesting point to note is in this section I use flows in bipartite networks liberally in the arguments in understanding the graph of the pedigree polytope (see Chap. 2 Sect. 2.3 for related definitions for a graph of a polytope in general). Some of the results proved can be obtained using other arguments, or known general results on adjacency of 0/1 polytopes, but, keeping in mind the use of flows in networks in Chaps. 5 and 6 as part of the framework developed, I stick to this approach. Thus, the machinery required is kept minimal, and the recursive definition of pedigrees is fully used.

Problem 4.3 (*Nonadjacency of Pedigrees*)
Instance:
  number of cities: n.
  pedigrees: $X^{[1]}, X^{[2]} \in P_n$.
Question: Are $X^{[1]}, X^{[2]}$ nonadjacent vertices of $conv(P_n)$?

Next, we give a formal definition of nonadjacency in pedigree polytopes imitating the general definition given in Chap. 2. Given $X^{[1]}, X^{[2]} \in P_n$, let $\bar{X} = \frac{1}{2}(X^{[1]} + X^{[2]})$. (In this section $\bar{X}$ has only this meaning unless otherwise specified.) Let $\lambda^0 \in \Lambda_{n-1}(\bar{X})$ correspond to the convex combination, $\bar{X}/n - 1 = \frac{1}{2}(X^{[1]}/n - 1 + X^{[2]}/n - 1)$. However, if $X^{[1]}/n - 1 = X^{[2]}/n - 1$, we define $\lambda^0(X^{[1]}/n - 1) = 1$.

Theorem 4.5 *Given vertices $X^{[i]} \in P_n$, $i = 1, 2$, they are nonadjacent in $conv(P_n)$ if and only if there exists a $S \subset P_n$ and $\mu \in \Lambda_n(\bar{X})$ such that*

- $S \cap \{X^{[i]}, i = 1, 2\} \neq \{X^{[i]}, i = 1, 2\}$,
- $\sum_{Y \in S} \mu(Y)Y = \bar{X}, \sum_{Y \in S} \mu(Y) = 1, \mu(Y) > 0, Y \in S$.

Such a S is called a witness for nonadjacency *of the given pedigrees, or* witness *for short.* ♡

Proof This result can be derived from the definition of the adjacency of vertices of a polytope, that is, the line segment joining the two vertices is an edge (one-dimensional face) of the polytope.   □

4.5.1 FAT Problems and Adjacency in Pedigree Polytope

Given $X^{[1]}, X^{[2]} \in P_n$, let the corresponding pedigrees be $W^{[i]}, i = 1, 2$. Let the $2 \times (n-3)$ array $L = (e_{ij})$ denote the edges in $W^{[1]}, W^{[2]}$ as rows, respectively. That is, $x_j^{[i]}(e_{ij}) = 1, i = 1, 2,$ and $4 \leq j \leq n$. We also informally say, e_{ij} is in $X^{[i]}$, if the corresponding edge is the ijth element of L.

> **Definition 4.7** Given $X/n - 1 \in conv(P_{n-1})$, consider any $\lambda \in \Lambda_{n-1}(X)$, and the $FAT_{n-1}(\lambda)$ problem. Then λ is called *inadmissible, rigid or flexible* depending on $FAT_{n-1}(\lambda)$ has no solution, unique solution or infinitely many solutions respectively. We also say that a λ is *admissible* if it is either rigid or flexible.   ♣

> **Lemma 4.7** *For any admissible* $\lambda \in \Lambda_{n-1}(X)$, *if problem* $FAT_{n-1}(\lambda)$ *has a single source or a single sink then* λ *is rigid.*   ♡

Proof This is so because if there is a single source/sink then the requirement/availability at any sink/source is the only feasible flow along the arc connecting the source/sink and the sink/source. In other words, the flow is unique and hence λ is rigid.   □

So, if $X^{[1]}/n - 1 = X^{[2]}/n - 1$ we have a single source in $FAT_{n-1}(\lambda)$ and so by Lemma 4.7, λ^0 is rigid. Similarly, if $x_n^{[i]}(e) = 1, i = 1, 2,$ for some $e \in E_{n-1}$ then $FAT_{n-1}(\lambda)$ has a single sink and any admissible λ is rigid. We have

> **Corollary 4.1** *If* $\lambda \in \Lambda_{n-1}(\bar{X})$ *is flexible then* $FAT_{n-1}(\lambda)$ *has at least two sources and exactly two sinks.*   ♡

Theorem 4.6 *Given* $X^{[1]}, X^{[2]} \in P_n$, $X^{[1]} \neq X^{[2]}$, *are adjacent in* $conv(P_n)$ $\Longleftrightarrow \lambda^0$ *is the only admissible* $\lambda \in \Lambda_{n-1}(\bar{X})$ *and it is rigid.* ♡

Proof Let $f(g, h)$ denote the flow along the arc (g, h) in $FAT_{n-1}(\lambda)$ corresponding to a feasible flow f. Consider $\lambda^0 \in \Lambda_{n-1}(\bar{X})$. Notice that the solution with $f(X^{[i]}/n - 1, e_{in}) = \frac{1}{2}, i = 1, 2$, and $f(g, h) = 0$, for all other arcs, is feasible for $FAT_{n-1}(\lambda^0)$. This implies λ^0 is admissible. Suppose λ^0 is the unique admissible $\lambda \in \Lambda_{n-1}(\bar{X})$ and λ^0 is rigid, we shall show that $X^{[1]}, X^{[2]}$ are adjacent in $conv(P_n)$.

λ^0 is rigid implies that the f given above is the only solution to $FAT_{n-1}(\lambda^0)$. This together with the fact λ^0 is the unique admissible $\lambda \in \Lambda_{n-1}(\bar{X})$ implies that the extensions of $X[i]/n - 1, i = 1, 2$, as per f are the only pedigrees in P_n which receive positive weights while representing $\bar{X}$. But the extensions are nothing but $X^{[1]}$ and $X^{[2]}$, this completes the proof one way.

Let $X^{[1]}, X^{[2]}$ be adjacent in $conv(P_n)$, and if possible let either (a) the unique admissible $\lambda \in \Lambda_{n-1}(\bar{X})$ be flexible, or (b) the set of admissible $\lambda \in \Lambda_{n-1}(\bar{X})$ be not a singleton set.

Case a: This implies λ^0, which is the unique admissible $\lambda \in \Lambda_{n-1}(\bar{X})$, is flexible. So, from an application of Corollary 4.1, $X[1]/n - 1 \neq X[2]/n - 1$ and similarly $e_{1n} \neq e_{2n}$. Notice that we have exactly two sources and two sinks in this problem. Flexibility of λ^0 means that there exists another solution $f' \neq f$ for $FAT_{n-1}(\lambda^0)$. Since $f(X^{[i]}/n - 1, e_{1n}) = \frac{1}{2}, i = 1, 2$; and $f(g, h) = 0$ otherwise, f', which differs from f must have $f'(X^{[i]}/n - 1, e_{in}) \neq \frac{1}{2}$ for some $i = 1, 2$, say, $i = 1$.

Now, $f'(X^{[1]}/n - 1, e_{1n}) = \theta$, is necessarily $< \frac{1}{2}$ as the availability at $X^{[1]}/n - 1$ is only $\lambda^0(X^{[1]}/n - 1) = \frac{1}{2}$. So the sink e_{1n} must get its remaining requirement $(\frac{1}{2} - \theta)$ from the other source, namely, $X^{[2]}/n - 1$. This further implies that the sink e_{2n} receives only θ from $X^{[2]}/n - 1$ and so it must receive its remaining requirement $(\frac{1}{2} - \theta)$, from the other source $X^{[1]}/n - 1$. Thus an alternative flow f' is possible for any $0 \leq \theta < \frac{1}{2}$ with

$$f'(X^{[1]}/n - 1, e_{1n}) = f'(X^{[2]}/n - 1, e_{2n}) = \theta,$$

$$f'(X^{[1]}/n - 1, e_{2n}) = f'(X^{[2]}/n - 1, e_{1n}) = \frac{1}{2} - \theta.$$

Thus, corresponding to $\theta = 0$, we have found a new set of pedigrees in P_n, that is, a witness for $X^{[1]}, X^{[2]}$ being nonadjacent in $conv(P_n)$. Contradiction. So (a) is not possible.

Case b: This implies, there exists an admissible $\lambda \in \Lambda_{n-1}(\bar{X}), \lambda \neq \lambda^0$, with a feasible flow f' for $FAT_{n-1}(\lambda)$.

So there exists a pedigree $Y \in P_{n-1}$, such that $\lambda(Y) > 0$ and $Y \neq X^{[1]}/n - 1$ (say, without loss generality). Let $f'(Y, e) > 0$ for some $e = e_{in}, i = 1, 2$. So corresponding to the admissible λ, and the feasible flow f', we have a $S \subset P_n$ such that

$S \neq \{X^{[1]}, X^{[2]}\}$ and $\bar{X}$ can be written as a convex combination of pedigrees in S. If $S \cap \{X^{[1]}, X^{[2]}\} \neq \{X^{[1]}, X^{[2]}\}$ we are through and $X^{[1]}, X^{[2]}$ are not adjacent in $conv(P_n)$. Otherwise let $S' = S - \{X^{[1]}, X^{[2]}\}$. Let $\mu \in \Lambda_n(\bar{X})$ be the weight vector corresponding to S. Let $min\{\mu(X^{[1]}), \mu(X^{[2]})\} = \epsilon$. Thus

$$\bar{X} = 1/(1 - 2\epsilon) \left\{ \sum_{Y \in S'} \mu(Y)Y + \sum_{i=1,2} (\mu(X^{[i]}) - \epsilon)X^{[i]} \right\}.$$

In other words, we have found a witness, which is a subset of S, that includes at most one of the two given pedigrees, $X^{[1]}$ and $X^{[2]}$. Contradiction. So (b) is not possible. Hence the theorem.   □

Lemma 4.8   *Given $X^{[1]}, X^{[2]} \in P_n$, suppose for some k, $4 \leq k < n$, and some $e \in E_k$, $x_{k+1}^{[i]}(e) = 1$, $i = 1, 2$, then every $\lambda \in \Lambda_k(\bar{X})$ is rigid.*   ♡

Proof Since $X^{[1]}, X^{[2]} \in P_n$, $X^{[1]}/k + 1$, $X^{[2]}/k + 1 \in P_{k+1}$. Therefore, $x_{k+1}^{[i]}(e) = 1$, $i = 1, 2$ for some $e \in E_k$, implies that $e \neq e_{il}, i = 1, 2$, for any l, $4 \leq l \leq k$. Also, there exists a l, such that, e_{il} is a generator of e, for $i = 1, 2$. (e_{1l}, e_{2l} may or may not be distinct).

Claim: For any $\lambda \in \Lambda_k(\bar{X})$, every $Y \in P_k$, with $\lambda(Y) > 0$, must agree with the zeros of $\bar{X}$. That is, $\forall l$, $4 \leq l \leq k$,

$$y_l = x_l^{[i]}, \text{ for some } i = 1, 2. \tag{4.25}$$

Proof Suppose, for some l if $y_l \neq x_l^{[i]}$ for both $i = 1, 2$ then $y_l(\bar{e}) = 1$ for some $\bar{e}, e_{1l} \neq \bar{e} \neq e_{2l}$. But $x_l(\bar{e}) = 0$. So, no such Y appears in any convex combination representing X/k. Hence the claim.   □

Consider any $\lambda \in \Lambda_k(X)$. Every $Y \in P_k$, with $\lambda(Y) > 0$ obeys Eq. (4.25). So, every Y has a generator of e. Thus we have the feasible flow f in $FAT_k(\lambda)$ given by $f(Y, e) = \lambda(Y)$. In other words λ is admissible. And the uniqueness of the flow implies λ is rigid as well. Hence the lemma.   □

As a consequence of Theorem 4.6, we have the following fact about *inheritance* of the adjacency property.

Theorem 4.7 *Given $X^{[1]}, X^{[2]} \in P_n$, suppose $X^{[1]}/k, X^{[2]}/k$ are adjacent / nonadjacent in $conv(P_k)$, for some $k, 4 \le k < n$, and $x_{k+1}^{[i]}(e) = 1$, $i = 1, 2$ for some $e \in E_k$, then $X^{[1]}/k + 1, X^{[2]}/k + 1$ are adjacent/nonadjacent in $conv(P_{k+1})$, accordingly.* ♡

Proof We have from Lemma 4.8, any $\lambda \in \Lambda_k(\bar{X})$ is rigid. Now $X^{[1]}/k, X^{[2]}/k$ are adjacent in $conv(P_k)$, implies we have a unique λ. And hence $\Lambda_{k+1}(\bar{X})$ is a singleton set. So $X^{[1]}/k + 1, X^{[2]}/k + 1$ are adjacent. And $X^{[1]}/k, X^{[2]}/k$ are nonadjacent in $conv(P_k)$, implies we have more than one $\lambda \in \Lambda_k(\bar{X})$. And from Lemma 4.8 all these λ's are rigid. And hence $\Lambda_{k+1}(\bar{X})$ has more than one element. Hence the inheritance of adjacency property is established. □

Remark (1) Theorem 4.7 can be repeatedly applied if $x_q^{[i]}(e_q) = 1$, $i = 1, 2$ for some $e_q \in E_{q-1}$, for all $k + 1 \le q \le s$, and we can conclude that $X^{[1]}/s, X^{[2]}/s$ are adjacent/nonadjacent in $conv(P_s)$ depending on $X^{[1]}/k, X^{[2]}/k$ are adjacent/nonadjacent in $conv(P_k)$.
(2) Notice that nothing certain can be said about inheritance of adjacency property, the moment we encounter a q with $\mathbf{x}_q^{[1]} \ne \mathbf{x}_q^{[2]}$. Examples 4.5 and 4.6 illustrate this point.
(3) Due to inheritance, it is sufficient to concentrate on components q such that $X^{[1]}$ and $X^{[2]}$ disagree. This leads to the definition of the set of discords.

Definition 4.8 Given $X^{[1]}, X^{[2]} \in P_n$, we call $D = \{q | \mathbf{x}_q^{[1]} \ne \mathbf{x}_q^{[2]}, 4 \le q \le n\}$ the *set of discordant components* or *discords*. This means, in terms of L, $e_{1q} \ne e_{2q}, q \in D$. ♣

Lemma 4.9 *Given $X^{[1]}, X^{[2]} \in P_n$, consider the set D. If $|D| = 1$ then $X^{[1]}$ and $X^{[2]}$ are adjacent in $conv(P_n)$.* ♡

Proof Let $q \in D$. We have $X^{[1]}/q - 1 = X^{[2]}/q - 1$ as q is the first component of discord. Let λ^0 correspond to the degenerate convex combination $\bar{X} = X^{[1]}/q - 1$. Consider $FAT_{q-1}(\lambda^0)$. From Lemma 4.7, λ^0 is rigid. So by Theorem 4.6 $X^{[1]}/q, X^{[2]}/q$ are adjacent in $conv(P_q)$. If $q = n$ we are through, otherwise from item 1 of the remark following Theorem 4.7, $X^{[1]}, X^{[2]}$ are adjacent in $conv(P_n)$. Hence the lemma. □

This lemma provides an easy-to-check sufficient condition for adjacency in the pedigree polytopes. So we have a nontrivial problem of determining nonadjacency, only when $|D| > 1$.

Lemma 4.10 *Given* $X^{[1]}, X^{[2]} \in P_n$, *consider the set* $D = \{q_1 < \ldots < q_r\}$. *Let* λ^0 *correspond to the convex combination* $X = \frac{1}{2}(X^{[1]}/q_r - 1 + X^{[2]}/q_r - 1)$. *If* $r > 1$ *and* λ^0 *is flexible then* $X^{[1]}$ *and* $X^{[2]}$ *are nonadjacent in* $conv(P_n)$.   ♡

Proof From an application of Theorem 4.6 we have $X^{[1]}/q_r$, $X^{[2]}/q_r$ nonadjacent in $conv(P_{q_r})$. And from the remark on inheritance following Theorem 4.7, $X^{[1]}, X^{[2]}$ are nonadjacent in $conv(P_n)$. Hence the lemma.   □

This lemma provides an easy-to-check sufficient condition for nonadjacency in the pedigree polytopes. So we have a nontrivial problem of determining nonadjacency, only when $|D| > 1$ and $FAT_{q_r-1}(\lambda^0)$ has an unique solution. This is equivalent to checking (recall the definition of the extension of a pedigree) whether

$$(X^{[\bar{i}]}/q_r - 1, \mathbf{x}^{[i]}_{q_r}) \text{ is a pedigree in } P_{q_r}, i = 1, 2, \tag{4.26}$$

where $\bar{i} = 3 - i$. This means a generator of e_{q_r} appears in $W^{[\bar{i}]}/q_r - 1$ and e_{q_r} itself does not appear in $W^{[\bar{i}]}/q_r - 1$.

Stated differently, failure to meet the conditions (4.26) can happen for two reasons: Either

(i) a *generator* of e_{iq_r} is *not available* in $W^{[\bar{i}]}/q_r - 1$. Or
(ii) e_{iq_r} *itself appears* in $W^{[\bar{i}]}/q_r - 1$, as it is used to insert some $l \leq q_r - 1$.

In case (i), in every $Y \in P_{q_r}$ which is eligible to be in any convex representation of $\bar{X}$ with a positive weight, we have $y_{q_r}(e_{iq_r}) = 1$ along with $y_l(e_{il}) = 1$ for some $l < q_r$ corresponding to a generator of e_{iq_r} available in $W^{[i]}/q_r - 1$. In case (ii), suppose $e_{\bar{i}l} = e_{iq_r}$, for some $l < q_r$, in every $Y \in P_{q_r}$ which is eligible to be in any convex representation of $\bar{X}$ with a positive weight, we have $y_{q_r}(e_{iq_r}) = 1$ along with $y_l(e_{il}) = 1$. Otherwise $y_l(e_{\bar{i}l}) = y_l(e_{iq_r})$ has to be 1. In that case, Y can not be a pedigree, as we have $y_{q_r}(e_{iq_r})$ also equal to 1.

Thus checking conditions (4.26) can be done using the procedure *find flexible* (Fig. 4.3). This involves at most $4n$ comparisons. But can be done more efficiently. We illustrate this procedure with Example 4.4.

Example 4.4 Consider the pedigrees in P_6, corresponding to

$$L = \begin{pmatrix} (1,2) & (2,4) & (2,5) \\ (1,3) & (1,2) & (2,3) \end{pmatrix}.$$

procedure *find flexible*
 1. **for** $i = 1, 2$
  begin
 2. $\bar{i} = 3 - i$
 3. Check whether $e_{iq_r} = e_{\bar{i}s}$, for some $s \leq q_r - 1$.
 4. **If** 'yes' **then** λ^0 is rigid, **break**;
 5. Check whether a generator e' of e_{iq_r}, is available in $W^{[\bar{i}]}/q_r - 1$.
 6. **If** 'no' **then** λ^0 is rigid, **break**;
 7. **if** $i = 2$ **then** λ^0 is flexible, **else continue**
  end
end *find flexible*.

Fig. 4.3 A procedure for finding whether λ^0 is flexible

So $D = (4, 5, 6)$ is the set of discords, with $q_r = 6$. In Step 1 of the procedure *find flexible*, we have $i = 1$, so $\bar{i} = 3 - i = 2$. In Step 3, we check whether $e_{16} = (2, 5) = e_{2s}$, for some $s \leq 5$. Since no such s exists we go to Step 5 and check whether a generator e' of e_{16}, is available in $W^{[2]}/5$. Since $e_{25} = (1, 2)$ is a generator of $(2, 5)$, the answer is yes, and we go to Step 7. As $i = 1$ we continue and go to Step 1. Now $i = 2$. And so $\bar{i} = 1$. In Step 3, we check whether $e_{26} = (2, 3) = e_{1s}$, for some $s \leq 5$. Since no such s exists we go to Step 5 and check whether a generator e' of e_{26}, is available in $W^{[1]}/5$. Since $e_{14} = (1, 2)$ is a generator of $(2, 3)$, the answer is yes and so in Step 7, we conclude that λ^0 is flexible and stop. From Lemma 4.10, we conclude that the given pedigrees are nonadjacent.

4.5.2 Graph of Rigidity and Its Implications

In this section, I define the graph of rigidity for a given pair of pedigrees, and show that the connectedness of the graph indicates that the pedigrees are adjacent. Otherwise, by swapping the edges in the pedigrees, corresponding to a component of the graph, we can produce a pair of new pedigrees, that forms a witness for the nonadjacency of the given pedigrees.

Definition 4.9 Let $\bar{i} = 3 - i$. Given L, giving the pedigrees $W^{[1]}, W^{[2]} \in \mathcal{P}_n$, a $q \in D$ and an $i \in \{1, 2\}$, we say that a generator of $e_{iq} = (u, v)$ is not available in $W^{[\bar{i}]}$ in case $e_{\bar{i}s} \notin G(e_{iq})$, where $s = max(4, v)$. Equivalently, $(W^{[\bar{i}]}/q - 1, e_{iq})$ is not a pedigree. ♣

Definition 4.10 Let $\bar{i} = 3 - i$. Given the pedigrees $W^{[i]}, i = 1, 2$. Let D be the set of discords. We say $q \in D$ is *welded to* $s, s \in D, s < q$ if either

(1) no generator of e_{iq} is available in the pedigree $W^{[\bar{i}]}$, for some $i = 1, 2$ *or*
(2) $e_{\bar{i}s} = e_{iq}$ for some $i = 1, 2$.   ♣

Definition 4.11 Given a pair of pedigrees in $\mathcal{P}_n$, we define the *graph of rigidity* denoted by G_R. The vertex set of G_R is the corresponding set of discords D, and the edge set is given by $\{(s, q) | s, q \in D, s < q,$ and q is welded to $s\}$.   ♣

Remark *(On G_R)*

(1) The graph G_R expresses the restriction imposed on the elements of D as far as producing a witness for nonadjacency of $X^{[1]}$ and $X^{[2]}$ in $conv(P_n)$ is concerned. Any $Y \in S \subset P_n$, a witness, has to agree with 0/1's of $\bar{X}$ and has to have exactly one edge from $\{e_{iq}, i = 1, 2\}, q \in D$. And so we may visualise Y as the incidence vector of a pedigree obtained from $X^{[1]}$ or $X^{[2]}$ by swapping (e_{1q}, e_{2q}), for some $q \in D$. (Definition 4.12 formalizes this idea.)
(2) Next we find conditions on G_R that will ensure nonadjacency of pedigrees. Notice that all q in a connected component of G_R are required to be *swapped* simultaneously, to ensure feasibility. Thus if G_R is a connected graph then we have no witness for nonadjacency, and so we can declare $X^{[1]}$ and $X^{[2]}$ are adjacent in $conv(P_n)$.

Definition 4.12 Given $C \subset V_n \setminus V_3$, let $Y^{[i]} = swap(X^{[i]}, C) \in B^{\tau_n}$ denote the characteristic vector, obtained from $X^{[i]}$ by swapping $q \in C$, where, by operation *swap* we mean:

$$y_q^{[i]} = \begin{cases} x_q^{[\bar{i}]} & \text{if } q \in C \\ x_q^{[i]} & \text{otherwise.} \end{cases}$$

Lemma 4.11 *Given* $X^{[i]} \in P_n$, $i = 1, 2$ *consider the graph* G_R. *If* $C = \{l\}$, *is a component of* G_R, *then* (i) $Y^{[i]}/l \in P_l$, $i = 1, 2$, *and so* (ii) $Y^{[i]} \in P_n$. ♡

Proof Suppose $Y^{[i]}/l \notin P_l$, for some $i = 1, 2$. That is, $Y^{[i]}/l = (X^{[i]}/l - 1, \mathbf{x}_l^{[\bar{i}]}) \notin P_l$. In other words there exists a $s < l$ such that l is welded to s and so (s, l) is an edge in G_R. But C is a component of G_R implies that $s \in C$. Contradiction. This proves part (i). If $l = n$ this also proves assertion (ii) of the lemma.

Let $l < n$. Suppose (ii) is false. Then there exists a q, $l < q \le n$, the smallest such, for which $Y^{[i]}/q \notin P_q$, for some $i = 1, 2$.

Notice that,

(1) $e_{iq} \ne e_{\bar{i}l}$, as otherwise q would be welded to l.
(2) $e_{iq} \ne e_{is}$, $s < q$, as otherwise that would contradict the fact $X^{[i]}/q$ is a pedigree in P_q.

Let $u < q$ be such that e_{iu} is the generator of e_{iq} in $X^{[i]}/q$.

Case 1: $u = l$. That is, e_{il} is the generator of e_{iq} in $X^{[i]}/q$. Since q is not welded to l, $e_{\bar{i}l}$ is also a generator of e_{iq}. And since $\mathbf{y}_l^{[i]} = \mathbf{x}_l^{[\bar{i}]}$, we have $e_{\bar{i}l}$ available in $Y^{[i]}/l$.

Case 2: $u \ne l$. Then e_{iu} is still available in $Y^{[i]}$ as $u \notin C$.

Thus we have shown that a generator of e_{iq} exists in $Y^{[i]}/q - 1$ and e_{iq} is not in $Y^{[i]}/q - 1$. Thus, $Y^{[i]}/q \in P_q$, $i = 1, 2$. Contradiction. This completes the proof of [ii].

Hence the lemma. □

Theorem 4.8 *Given* $X^{[i]} \in P_n$, $i = 1, 2$, *if* C *is a component of* G_R, *then* $Y^{[i]} = swap(X^{[i]}, C) \in P_n$, $i = 1, 2$. ♡

Proof We prove this theorem by induction on the cardinality of C and on the cardinality of D, the vertex set of G_R. Lemma 4.11 provides the basis for induction. Suppose the theorem is true for any component with cardinality up to $r - 1$, and the set of discards having up to $s - 1$ vertices.

Let $C = \{l_1 < l_2 \ldots < l_r\} \subset D$, be a component of G_R. Consider the graph of rigidity G_R' with vertex set $D \setminus \{l_r\}$. Now $C \setminus \{l_r\}$ in G_R' may or may not be a single component of G_R'. Consider $Y^{[i]}/l_r - 1$, $i = 1, 2$ obtained by swapping $C \setminus \{l_r\}$ in $X^{[i]}/l_r - 1$, $i = 1, 2$. Now $Y^{[i]}/l_r - 1 \in P_{l_r-1}$, $i = 1, 2$ by induction hypothesis. Since l_r is welded to some element(s) in $C \setminus \{l_r\}$ in G_R, $(Y^{[i]}/l_r - 1, \mathbf{x}_{l_r}^{[\bar{i}]}) \in P_{l_r}$, $i = 1, 2$. Which is equivalent to swapping C in $X^{[i]}$, $i = 1, 2$. Therefore, $Y^{[i]}/l_r \in P_{l_r}$. The proof of $Y^{[i]} \in P_n$, $i = 1, 2$ is similar to that of Lemma 4.11. Hence the theorem. □

4.5.3 Characterisation of Nonadjacency Through the Graph of Rigidity

From the results obtained on the graph of rigidity, we are in a position to interpret the nonadjacency of a given pair of pedigrees, using the graph G_R. We have shown that for any component C of G_R, $swap(X^{[i]}, C)$ produces a pedigree in P_n. However if $C = D$ then the swapping produces, trivially, the same pedigrees, as $swap(X^{[i]}, D)$ is $X^{[\bar{i}]}$. So if $C \neq D$ we get a pair of pedigrees from $X^{[i]}, i = 1, 2$, by swapping C and it is easy to check that, we have a witness for nonadjacency of the given pedigrees, that is, $\bar{X} = \frac{1}{2}(Y^{[1]} + Y^{[2]})$ and $X^{[i]}, Y^{[i]}, i = 1, 2$, are all different. Thus we have the following theorem characterising nonadjacency in pedigree polytopes.

> **Theorem 4.9** *Given $X^{[i]} \in P_n, i = 1, 2$, consider the graph of rigidity G_R. The given pedigrees are nonadjacent in $conv(P_n)$ if and only if G_R is not connected.*   ♡

Given two pedigrees, the set of discords, D, can be found in at most $n - 3$ comparisons. If $|D| = 1$ we stop as the pedigrees are adjacent. Otherwise, $D = \{q_1 < \cdots < q_r\}$. Construction of G_R requires finding the edges in G_R, which can be done starting with q_r and checking whether it is welded to any $s < q_r, s \in D$. Something similar to the procedure *find flexible* is required. At most $4n$ comparisons are required, as noted earlier. Then we do the same with q_{r-1} and so on. So the construction of G_R is of complexity $O(n^2)$. However, it is well known that the components of an undirected graph can be found efficiently, $O(|V| + |E|)$ [2]. Thus we have an algorithm that can check nonadjacency in the pedigree polytope, $conv(P_n)$ in time strongly polynomial in n. The following examples illustrate the simple algorithm.

Example 4.5 Consider the pedigrees in P_6, corresponding to

$$L = \begin{pmatrix} (1, 2) \ (2, 3) \ (2, 5) \\ (1, 2) \ (2, 4) \ (2, 3) \end{pmatrix}.$$

So $D = \{5, 6\}$ is the set of discords. Here, 6 is welded to 5 as $e_{26} = (2, 3) = e_{15}$. So $G_R = (\{5, 6\}, \{(5, 6)\})$.

Since G_R has a single component, from Theorem 4.9, the pedigrees are adjacent in $conv(P_6)$.

Example 4.6 Consider the pedigrees in P_6, corresponding to

$$L = \begin{pmatrix} (1, 3) \ (2, 3) \ (3, 4) \\ (1, 2) \ (1, 4) \ (1, 3) \end{pmatrix}.$$

So $D = \{4, 5, 6\}$. Here, 6 is welded to 4 for more than one reason. ($e_{26} = (1, 3) = e_{14}$ and no generator of $e_{16} = (3, 4)$ is in the second pedigree.) But 5 is not welded to any other element of D. So $G_R = (\{4, 5, 6\}, \{(4, 6)\})$.

Since G_R has two components, $C_1 = \{4, 6\}$ & $C_2 = \{5\}$; from Theorem 4.9, the pedigrees are nonadjacent. The new set of pedigrees $\{((1, 3), (1, 4), (3, 4)), ((1, 2), (2, 3), (1, 3)\}$, obtained by swapping the component, $C = \{5\}$ is a witness.

Example 4.7 Consider the pedigrees in P_7, corresponding to

$$L = \begin{pmatrix} (1, 2) \ (2, 4) \ (2, 3) \ (2, 6) \\ (1, 3) \ (2, 3) \ (2, 5) \ (3, 4) \end{pmatrix}.$$

So $D = \{4, 5, 6, 7\}$. Here 7 is welded to 4 as no generator of $e_{27} = (3, 4)$ is in the first pedigree. But 6 is welded to 5 as $e_{16} = (2, 3) = e_{25}$. Finally, 5 is welded to 4 as no generator of $e_{15} = (2, 4)$ is in the second pedigree. So $G_R = (\{4, 5, 6, 7\}, \{(4, 5), (4, 7), (5, 6)\})$.

G_R is connected and can be easily seen. So from Theorem 4.9, the pedigrees are adjacent.

4.5.4 Adjacency in Pedigree Polytope Does Not Imply Adjacency in Tour Polytope

The examples provided in the previous section have something in common, namely, whenever the pedigrees are adjacent/nonadjacent in the $conv(P_n)$ the corresponding n-tours are also adjacent/ nonadjacent in Q_n. If this were in general true that would imply NP = P since we have a one-to-one correspondence between pedigrees and tours. Therefore, we are interested in the question: Does adjacency/nonadjacency of a pair of pedigrees in $conv(P_n)$ imply adjacency/nonadjacency of the corresponding n-tours in Q_n? The answer is negative. We give a counter example to show that the adjacency of pedigrees does not imply adjacency of the corresponding n-tours. However, we shall show in Sect. 4.6 that nonadjacency of pedigrees implies nonadjacency of corresponding tours in the tour polytope.

Example 4.8 Consider the pedigrees $W^{[1]}$, $W^{[2]}$ in P_{10}, corresponding to

$$L = \begin{pmatrix} (1, 2) \ (1, 3) \ (2, 4) \ (2, 6) \ (3, 5) \ (1, 4) \ (5, 8) \\ (1, 3) \ (2, 3) \ (3, 4) \ (4, 6) \ (3, 5) \ (1, 4) \ (4, 7) \end{pmatrix}$$

So $D = (4, 5, 6, 7, 10)$. Here 10 is welded to 7 as no generator of $e_{210} = (4, 7)$ is in the first pedigree. But 7 is welded to 6 as no generator of $e_{17} = (2, 6)$ is in the second pedigree. 6 is welded to 4, as no generator of $e_{26} = (3, 4)$ is in the first pedigree. And finally, 5 is welded to 4 as $e_{15} = (1, 3) = e_{24}$. So $G_R = (\{4, 5, 6, 7, 10\}, \{(4, 5), (4, 6), (6, 7), (7, 10)\})$. As G_R is connected, from Theorem 4.9, the pedigrees $W^{[1]}$, $W^{[2]}$ are adjacent in $conv(P_n)$.

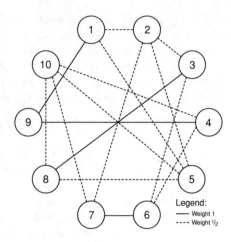

Let the corresponding 10-tours be called $Tour_1$, $Tour_2$ respectively. Let the inci-
dent vector corresponding to $Tour_i$ be denoted by $T^{[i]}$. Let $\overline{T} = \frac{1}{2}[T^{[1]} + T^{[2]}]$. We
have

$$Tour_1 = (1, 9, 4, 6, 7, 2, 3, 8, 10, 5, 1) \text{ and } Tour_2 = (1, 9, 4, 10, 7, 6, 3, 8, 5, 2, 1).$$

Now consider the tours $Tour_3$, $Tour_4$ given by,

$$Tour_3 = (1, 9, 4, 10, 8, 3, 6, 7, 2, 5, 1) \text{ and } Tour_4 = (1, 9, 4, 6, 7, 10, 5, 8, 3, 2, 1).$$

Let $\overline{T}' = \frac{1}{2}[T^{[3]} + T^{[4]}]$. It can be verified that $\overline{T} = \overline{T}'$. In other words, we have
shown that $T^{[1]}$, $T^{[2]}$ are nonadjacent in Q_{10}. *Figure 4.4 gives the support graph
of $\overline{T}$.*

4.6   Nonadjacency in Pedigree Polytope Implies
Nonadjacency in Tour Polytope

Next, we shall show that the nonadjacency of pedigrees in $conv(P_n)$ indeed implies
the corresponding tours are nonadjacent in Q_n the *STSP* polytope.

First, we prove the following claim for a polytope given by a set of inequalities
and then apply the same to the polytope $P_{MI}(n)$, to prove the main result.

Given $A, D \in R^{m \times n}$, $b, d \in R^m$, consider the bounded feasible region, $\{x \in R^n | Ax \le b, Dx = d, x \ge 0\}$. Let $\{x^s, s \in S\}$ be the set of extreme points of the fea-
sible region. Let u be the slack variable vector such that $F = \{(x, u) \in R^{m+n} | Ax + Iu = b, Dx = d, x \ge 0, u \ge 0\}$. Notice that u is unique for a given $x^s \in S$. Not only
that, u is unique for any x in the feasible region as well.

Claim: Assuming the above, for any $x^s \in S$ let u^s be $\ni Ax^s + Iu^s = b$. Suppose $x \in F$ is $\ni x = \sum_{i \in J} \lambda_i x^i$, $\sum_{i \in J} \lambda_i = 1$ and $\lambda_i \geq 0$, for some $J \subseteq S$ of extreme points, then the unique u corresponding to x can be written as

$$u = \sum_{i \in J} \lambda_i u^i.$$

Proof $(x^i, u^i) \in F$, $\forall i \in J$. Therefore $Ax^i + Iu^i = b$, $Dx^i = d$, $\forall i \in J$. Hence $Ax = A(\sum_{i \in J} \lambda_i x^i) = \sum_{i \in J} \lambda_i Ax^i$. But $Ax^i = b - u^i$. So we have $Ax = \sum_{i \in J} \lambda_i (b - u^i) = \sum_{i \in J} \lambda_i b - \sum_{I \in J} \lambda_i u^i = b - \sum_{i \in J} \lambda_i u^i$. But $Ax = b - u$ as well. Hence the result. □

Remark (i) Even if x can be expressed using two different convex combinations of subsets of extreme points, say, J and J', the slack variable vector u remains the same. However, we can represent u in different ways using λ's corresponding to extreme points x^is in J or J' respectively.

(ii) In general, given $u \geq 0$ we cannot claim there is a unique $x \ni (x, u) \in F$ (see Example 4.9).

(iii) Equality constraints do not play any restrictive role in this result.

Example 4.9 Consider a *STSP* with $n = 5$. Let X be given by $x_4(1, 2) = x_4(1, 3) = x_5(1, 3) = x_5(2, 4) = \frac{1}{2}$ with corresponding slack variable vector u given by $u_{12} = \frac{1}{2}$, $u_{13} = 0$, $u_{23} = 1$, $u_{14} = 1$, $u_{24} = 0$, $u_{34} = \frac{1}{2}$, $u_{15} = u_{25} = u_{35} = u_{45} = \frac{1}{2}$. It can be verified that we have a fractional basic feasible solution to the MI-relaxation problem, and the basis inverse is given by Table 4.1.

Now observe that the fractional solution given above can be written as a convex combination of the two integer solutions to the MI-relaxation given by

$$x_4(1, 3) = x_5(1, 2) = u_{23} = u_{14} = u_{34} = u_{15} = u_{25} = 1,$$

and

$$x_4(1, 3) = x_5(3, 4) = u_{12} = u_{23} = u_{14} = u_{35} = u_{45} = 1$$

with weights, $(\frac{1}{2}, \frac{1}{2})$.

Since the slack variable vectors above are tours, the slack variable vector corresponding to a fractional basic solution to the MI-formulation is in the convex hull of tours. Figure 4.5 gives the support graph of the u vector.

Theorem 4.10 *Let $T(X)$ denote the tour corresponding to the pedigree X. Consider $X^{[1]}, X^{[2]} \in P_n$ such that they are nonadjacent pedigrees in $conv(P_n)$. Then $T(X^{[1]})$, $T(X^{[2]})$ are nonadjacent in Q_n, the STSP polytope.* ♡

Table 4.1 Inverse of the basis matrix for Example 4.9

Basic variables:

$x_4(1, 2), x_4(1, 3), x_5(2, 4), x_5(1, 3), u_{12}, u_{23}, u_{14}, u_{34}, u_{15}, u_{25}, u_{35},$ and u_{45}

$$
\begin{pmatrix}
1/2 & -1/2 & 0 & 1/2 & 0 & 0 & 1/2 & 0 & 0 & 0 & 0 & 0 \\
-1/2 & 1/2 & 0 & 1/2 & 0 & 0 & -1/2 & 0 & 0 & 0 & 0 & 0 \\
1/2 & 1/2 & 0 & -1/2 & 0 & 0 & 1/2 & 0 & 0 & 0 & 0 & 0 \\
-1/2 & -1/2 & 1 & 1/2 & 0 & 0 & 1/2 & 0 & 0 & 0 & 0 & 0 \\
0 & 0 & 0 & 0 & 1 & 0 & 0 & 0 & 0 & 0 & 0 & 0 \\
1 & 0 & 0 & 0 & 0 & 1 & 0 & 0 & 0 & 0 & 0 & 0 \\
1/2 & -1/2 & 0 & 1/2 & 0 & 0 & 1/2 & 1 & 0 & 0 & 0 & 0 \\
-1/2 & 1/2 & 0 & 1/2 & 0 & 0 & -1/2 & 0 & 1 & 0 & 0 & 0 \\
1/2 & 1/2 & 0 & -1/2 & 0 & 0 & 1/2 & 0 & 0 & 1 & 0 & 0 \\
-1/2 & 1/2 & 0 & 1/2 & 0 & 0 & -1/2 & 0 & 0 & 0 & 1 & 0 \\
1/2 & 1/2 & 0 & -1/2 & 0 & 0 & 1/2 & 0 & 0 & 0 & 0 & 1
\end{pmatrix}
$$

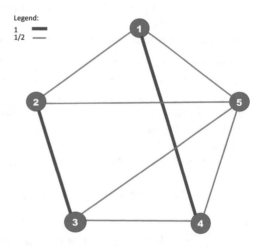

Fig. 4.5 Support graph for slack variable vector u in Example 4.9

Legend:
1 ▬
1/2 ▭

Proof Consider $X^{[1]}, X^{[2]}$ nonadjacent in $conv(P_n)$.

From Lemma 3.1 we know that the tour $T(X)$ corresponding to any pedigree X is the slack variable vector corresponding to X, and $(X, T(X))$ is feasible for the MI-relaxation problem.

Now consider $\bar{X} = \frac{1}{2}(X^{[1]} + X^{[2]})$. $\bar{X}$ is a feasible solution to the MI-relaxation problem as $X^{[1]}, X^{[2]}$ are pedigrees and $\bar{X}$ is the midpoint of the line segment joining them. Let U be the slack variable vector corresponding to $\bar{X}$.

Now apply the claim made earlier in this section with $\bar{X}$ and its representation using $X^{[1]}, X^{[2]}$. Thus we have $U = \frac{1}{2}(T(X^{[1]}) + T(X^{[2]}))$.

Since $X^{[1]}, X^{[2]}$ are nonadjacent in $conv(P_n)$ there exists two other pedigrees $Y^{[1]}, Y^{[2]}$ such that $X^{[1]} + X^{[2]} = Y^{[1]} + Y^{[2]}$. So $\bar{X}$ is also equal to $\frac{1}{2}(Y^{[1]} + Y^{[2]})$.

Corresponding to the pedigree $Y^{[i]}$, $T(Y^{[i]})$ is the slack variable vector in the MI-formulation, $i = 1, 2$.

Now apply the claim made above once more with $\bar{X}$, its representation using $Y^{[1]}$, $Y^{[2]}$ and the slack variable vector U. We have $U = \frac{1}{2}(T(Y^{[1]}) + T(Y^{[2]}))$.

Therefore we have expressed U in two different ways using four different tours. That is, we have shown that $T(X^{[1]}) + T(X^{[2]}) = T(Y^{[1]}) + T(Y^{[2]})$. This implies the tours $T(X^{[1]})$, $T(X^{[2]})$ are nonadjacent in the $STSP$ polytope, Q_n. Hence the theorem. □

Remark • In this proof, we have made explicit use of the properties of the MI-relaxation and the pedigree polytope.

• Theorem 4.10 provides a sufficient condition for nonadjacency of tours in Q_n.
• Notice that this sufficient condition can be checked efficiently as checking nonadjacency in the pedigree polytope can be done efficiently as shown in Sect. 4.5.3.

4.6.1 Pedigree Polytope is a Combinatorial Polytope

The main result of this subsection is the observation that the pedigree polytopes are combinatorial polytopes. The implications of this and other results relating to some of the properties listed earlier in Sect. 2.6.1 are also discussed.

In this chapter, in Sect. 4.5.3, we defined the graph of rigidity for a given pair of pedigrees. And showed that the connectedness of the graph answers the adjacency/nonadjacency decision question. The graph G_R expresses the restriction imposed on the elements of the set of discards, D as far as producing a witness for nonadjacency of $X^{[1]}$ and $X^{[2]}$ in $conv(P_n)$ is concerned. Any $Y \in S \subset P_n$, a witness, has to agree with the components on which both $X^{[1]}$, $X^{[2]}$ themselves agree and has to have exactly one edge from $\{e_{iq}, i = 1, 2\}$, $q \in D$. And so we may visualise Y as the incidence vector of a pedigree obtained from $X^{[1]}$ or $X^{[2]}$ by swapping (e_{1q}, e_{2q}), for some $q \in D$. Using Theorem 4.8 we can conclude that if G_R is not a connected graph, then the set of vertices in C of any connected component of G_R is a proper subset of D and swapping them produces a set $S = \{Y^{[1]}, Y^{[2]}\} \subset P_n$. And S is a witness for nonadjacency of the given $X^{[1]}$, $X^{[2]}$ in $conv(P_n)$. Notice that all q in a connected component of G_R are required to be *swapped* simultaneously, to obtain a legitimate swap. That is, each connected component of G_R is minimally legitimate, producing two other pedigrees, $Y^{[1]}$, $Y^{[2]}$ as a witness. Thus, we have shown that $conv(P_n)$ is a combinatorial polytope. We state this fact as a Theorem 4.11. We have outlined earlier in Sect. 4.5.3, a strongly polynomial algorithm to test whether a given pair of pedigrees are nonadjacent.

Theorem 4.11 *The pedigree polytope, $conv(P_n)$ is a combinatorial polytope.*
♡

4.6.1.1 Other Properties Shared by Pedigree Polytopes

Recall that several properties of polytopes were listed in Sect. 2.6.1 and their implications and inclusions proved in the literature were summarised at the end of that
subsection. Here we study which among them are satisfied by the pedigree polytopes.

Matsui and Tamura [120] give the following counter example to show that combinatorial property (Definition 2.37) does not imply property B (Definition 2.36):

Example 4.10 Consider the set of $0 - 1$ vectors $V = \{(0, 0), (1, 0), (1, 1)\}$.
$conv(V)$ is combinatorial in a vacuous sense as the vertices are all mutually adjacent
in $conv(V)$. Though $((0, 0), (1, 0), (1, 1))$ is a monotone vertex sequence, the vector
$(0, 0) - (1, 0) + (1, 1) = (0, 1) \notin V$. So it does not satisfy the property B.

Notice that for $n = 4$, $P_4 = \{(1, 0, 0), (0, 1, 0), (0, 0, 1)\}$. The vertices are mutually adjacent. And there is no monotone vertex sequence available here. So in a
vacuous sense $conv(P_4)$ has property B.

However, we shall show that the pedigree polytopes for $n > 4$ do not possess
property B.

Theorem 4.12 *The pedigree polytopes, $conv(P_n)$, $n > 4$ do not satisfy Matsui
and Tamura's Property B (Definition 2.36).*   ♡

Proof We prove the result by induction on n. The basis for induction is provided
by $conv(P_5)$. Consider the vertex sequence $\rho = (X^{[1]}, (X^{[2]}, X^{[3]})$ corresponding
to the pedigrees $((1, 2)(1, 3)), ((1, 2)(1, 4)), ((1, 3)(1, 4))$ respectively. It is easy
to check that ρ is a monotone vertex sequence (using Definition 2.34). But $X^{[1]} -
X^{[2]} + X^{[3]} = (0, 1, 0, 0, 1, 0, 0, 0, 0, 0)$ corresponds to $((1, 3), (1, 3))$ and it is not a
pedigree as it violates the requirement that the edges in a pedigree are distinct. Hence
$conv(P_5)$ does not have property B.

Now assume that property B is violated by all polytopes $conv(P_k)$, for $5 \leq k \leq
n - 1$. Now we shall show that so is the case with $conv(P_n)$.

From the induction hypothesis $\exists$ a monotone vertex sequence $\rho = (X^{[1]}, X^{[2]},
X^{[3]})$ in $conv(P_{n-1})$ such that $X^{[1]} - X^{[2]} + X^{[3]}$ is not a vertex of $conv(P_{n-1})$. Now
consider the last component of these vectors in the sequence, namely, $\mathbf{x}_{n-1}^{[i]}$, $i = 1, 2$
and 3 respectively.

Consider the edge $e = (u, v)$ for which $\mathbf{x}_{n-1}^{[2]}(e) = 1$. As ρ is a monotone vertex
sequence, we have three cases to consider.

Case 1: $\mathbf{x}_{n-1}^{[1]}(e) = 1$ and $\mathbf{x}_{n-1}^{[3]}(e) = 1$. Consider $e' = (u, n - 1)$ let $\mathbf{y}(e')$ be the
indicator of e'. Notice that $(X^{[i]}, \mathbf{y}(e'))$ is a pedigree in P_n for all $1 \leq i \leq 3$. And
the vertex sequence $\rho' = ((X^{[1]}, \mathbf{y}(e')), (X^{[2]}, \mathbf{y}(e')), (X^{[3]}, \mathbf{y}(e')))$ in $conv(P_n)$ is
monotone. But fails to satisfy property B.

Case 2: $x_{n-1}^{[1]}(e) = 1$ and $x_{n-1}^{[3]}(e) = 0$. Let $e' = (u, n - 1)$. Notice that $(X^{[i]}, y(e'))$ is a pedigree in P_n for $i = 1, 2$. Choose e'' such that $(X^{[3]}, y(e''))$ corresponds to a pedigree in P_n. Now the vertex sequence $\rho' = ((X^{[1]}, y(e')), (X^{[2]}, y(e')), (X^{[3]}, y(e'')))$ in $conv(P_n)$ is monotone. But fails to satisfy property B.

Case 3: $x_{n-1}^{[1]}(e) = 0$ and $x_{n-1}^{[3]}(e) = 1$. The proof is similar to *Case 2*.

Hence $conv(P_n)$ does not have property B. $\qquad\square$

So we have non-vacuous examples of combinatorial polytopes, which do not satisfy property B stipulated by Matsui and Tamura.

The fact that the pedigree polytopes are combinatorial can be used to assert the strong adjacency property (Definition 2.40) for the pedigree polytopes. Some of these properties are used in later chapters. However, other properties of pedigree polytopes can be useful in designing newer approaches to solve problems related to pedigree polytopes.

4.6.2 Diameter of the Pedigree Polytope and Related Results

As we noticed in Chap. 2, the diameter of a polytope is studied, for its own sake as an interesting property of polytopes, in addition to the hope of designing algorithms that could exploit the improvement along the vertex transitions. M. Padberg and M. R. Rao [137] showed that the diameter of the polytope associated with the asymmetric travelling salesman problem as well as the diameter of the polytopes associated with a large class of combinatorial problems that includes well-solved problems such as the assignment problem and the matching problem is less than or equal to two.

This result has prompted them to conjecture:

...our result seems to indicate that there may exist "good" algorithms for a large class of problems. This includes not only the travelling salesman problem, but also - as Karp [101] has noted - a much larger class of classic but as of yet not satisfactorily solved combinatorial problems including the general zero-one programming problem.

As mentioned earlier in Chap. 2, Rispoli and Cosares [149] showed that the diameter of *STSP* polytope is at most 4, for every $n \geq 3$, and is thus independent of n. What can we say about the diameter of Pedigree Polytope $conv(P_n)$?

For $n = 4$, $conv(P_n)$ has diameter 1. Notice that given any $P^1 \in conv(P_5)$, there exists exactly one $P^2 \in conv(P_5)$ that is nonadjacent, (see Fig. 6.1). Therefore, the diameter of the pedigree polytope $conv(P_5)$ is 2. In Sect. 4.6, we showed that the nonadjacency of two pedigrees implies the nonadjacency of the corresponding tours in Q_n, the *STSP* polytope. However, any two pedigrees that are adjacent in the pedigree polytope may be such that the corresponding tours are nonadjacent, as shown by Example 4.8.

Suppose that for some n, $conv(P_n)$ has a diameter greater than the diameter of Q_n. Then we have nonadjacent pedigrees $P^1, P^2 \in conv(P_n)$ such that the minimum number of edges in a path joining them in $G(conv(P_n))$ the graph of $conv(P_n)$ is greater than the diameter of Q_n. Consider the corresponding tours of $T(P^i) \in Q_n$, $i = 1, 2$. They are nonadjacent. There is a path joining $T(P^1)$ and $T(P^2)$ in the graph of Q_n with the number of edges less than or equal to the diameter of Q_n. Each edge in this path connects adjacent tours, so the corresponding pedigrees are adjacent in $G(conv(P_n))$. Thus we have an edge path connecting P^1, P^2 with a length less than or equal to the diameter of Q_n, contradicting our supposition. Hence, we can state this fact as Theorem 4.13.

Theorem 4.13 *The diameter of $conv(P_n)$ is less than or equal to the diameter of $Q_n \le 4$.*   ♡

What more do we know about the diameter of the pedigree polytopes? Mozhgan Pourmorradnasseri in her doctoral thesis [142] discusses the asymptotic behaviour of the graph of the pedigree polytope, from her research collaborating with Abdullah Makkeh, and Dirk Oliver Theis [116, 117]. In [116] the authors use the graph of rigidity given by Definition 4.11, for two pedigrees in P_n, and work with the corresponding tours A, B respectively, and denote it as G_n^{AB} making it explicitly depend on A, B the specific tours and n. Here, tour A is chosen by Alice, according to a strategy and tour B is chosen at random by Bob blindfolded. And the game continues as the players choose their respective n-tours (or equivalent pedigrees from P_n), $n \to \infty$. Consider two tours A and B, that is, Alice's tour and Bob's tour, on the vertex set $[n]$. At the time $n + 1$, Alice and Bob insert the vertex $n + 1$ to their respective tours chosen in the previous step. There are three eventualities possible: [1] The pedigree graph G_{n+1}^{AB} stays the same as G_n^{AB} or [2] it modifies G_n^{AB} by adding the isolated vertex $n + 1$ or it modifies G_n^{AB} by adding the vertex $n + 1$ together with edges connecting $n + 1$ and some vertices in $[n]$.

Event [1] happens if Alice and Bob choose the same edge to insert $n + 1$, in that case, there is no discard and so $n + 1$ is not a node in the rigid graph. Event [2] results if the edges chosen by Alice and Bob are different, but $n + 1$ is not welded to any of the other nodes in the set of discards. Finally, event [3] happens if the edges chosen for insertion by Alice and Bob are different, but $n + 1$ is welded to some other nodes in the set of discards.

They use an adjacency game in which Alice wins if the pedigrees A and B are eventually nonadjacent, Bob wins otherwise. Notice that the event [2] discussed above contributes to a disconnected graph of rigidity. Therefore, the expected value of the random variable, Y, that counts the number of times the event [2] happens, plays a crucial role in their proof of their main result, rephrased here as Theorem 4.14.

Theorem 4.14 *For every $\varepsilon > 0$ there is an integer N such that for all $n \geqslant N$ and all tours A_n with node set $[n]$, if B_n is drawn uniformly at random from all tours with node set $[n]$, then*

$$\mathbb{P}\,(\text{The graph of rigidity corresponding to } A_n \text{ and } B_n \text{ is connected}\,) \geqslant 1 - \varepsilon.$$

In other words, using infinite tours, this reads:

$$\forall \varepsilon > 0 \quad \exists N : \forall A, \quad \forall n \geqslant N : \mathbb{P}\left(G_n^{AB} \text{ is connected}\right) \geqslant 1 - \varepsilon,$$

$\heartsuit$

This theorem is equivalent to saying, the minimum degree is asymptotically equal to the number of vertices, $(n-1)!/2$ i.e., the graph is "asymptotically almost complete". The authors obtain this asymptotic result on the graph of $conv(P_n)$, using Markov decision process-like models analysing the adjacency game (see [116] for proofs, and related concepts and descriptions).

4.6.3 What Next?

The next three chapters are very crucial in our strategy to invent an analogy, framework and implementable algorithm based on what has been learnt so far on pedigrees. Dynamic programming formulations of the *STSP* problem (Bellman [28], Held and Karp [91]) can be viewed as algorithms based on the $\mathcal{V}$-presentation of *STSP* polytope. The standard formulation of the *STSP* given by [57] initiates the era of attacking the *STSP* using a partial description of the $\mathcal{H}$-presentation of the polytope. This description though started with exponentially many facets of the polytope, till date we have not obtained an irredundant description of the polytope except for very small n. However, in general, several classes of facet-defining inequalities have been identified by researchers for *STSP* polytope and several other *COP*s. For *STSP* we see Martin Grötschel, and Manfred Padberg's articles [84, 85] discuss other inequalities (comb inequalities) than the ones defining *SEP(n)*. As listed in [112], "Several generalisations of the comb inequalities are known, with exotic names like clique tree, star, hyperstar, bipartition and binested inequalities. There are also several other classes of facet-inducing inequalities known, such as chain, ladder, crown, and hypohamiltonian inequalities."

Despite the tremendous success of the branch-and-cut approach for *STSP* and other *COP*'s, one thing we notice is we are adding classes of inequalities (facets) discovered in $\mathbb{Q}^{((n(n-3)/2)-1)}$, involving variables defined on the edges of K_n. We will deviate from the strategies to identify the facets of the pedigree polytope, other than the ones defining *MI*-relaxation polytope.

Then how do we propose a method to solve the pedigree optimisation problem? We solve the membership problem, that is, given an $X \in \mathbb{Q}^{T_n}$ we check whether X is in the pedigree polytope, $conv(P_n)$.

Here comes a very interesting analogy from biology. Let us look at proteins. The following quote from an interview with biologist Nathan Ahlgren, in *Conversation*[1] provides a simple description of a protein: "A protein is a basic structure that is found in all life. It's a molecule. And the key thing about a protein is it's made up of smaller components, called amino acids. I like to think of them as *a string of different coloured beads*. Each bead would represent an amino acid, which are smaller molecules containing carbon, oxygen, hydrogen, and sometimes sulphur atoms. Thus, a protein essentially is a string that is made up of these little individual amino acids. A protein doesn't usually exist as a string, but actually folds up into a particular shape, depending on the order and how those different amino acids interact together."

The analogy is to look at a pedigree, a string of 0/1 in the higher dimension, namely, $\mathbb{Z}^{T_n}$, as a flow through a folded two-dimensional layered network. This network has $n - 2$ layers; We have a layer for k, $k \in [4, n]$, having p_{k-1} nodes, with capacity x_{ijk} on node (i, j, k). The details of which will be the subject matter for Chap. 5. A rough interpretation is as follows:

Pedigrees	Interpretation in	
	Symbol	Proteins
cities	$i, j, k \in [n]$	atoms
triangles	$\{i, j, k\}$	amino acids
common edge	(i, j)	bonding between amino acids
capacity on node $\{i, j, k\}$	x_{ijk}	potential of an amino acid to transport
multi-layered network in $2 - D$	N_k	folding in $3 - D$
flow through the network	f	flow of different vital ingredients
membership in pedigree polytope	$X \in conv(P_n)$	recognition of different proteins by shape

---

[1] What is protein?, Nathan Ahlgren, appeared in Conversation, https://theconversation.com/what-is-a-protein-a-biologist-explains-152870.

Unfortunately, we do not have any available theory developed for such a layered network interpretation of pedigrees, so I provide it in Chap. 5. And based on that theory, a framework for solving the membership problem for pedigree polytopes is offered.

Chapter 5
Membership Checking in Pedigree Polytopes

5.1 Introduction

This chapter is devoted to the membership problem of pedigree polytopes. As explained in Sect. 4.6.3, we are given $X \in \mathbb{Q}^{\tau_n}$, we wish to check $X \in conv(P_n)$. Necessarily, $X \in P_{MI}(n)$, otherwise, we can straight away conclude $X \notin conv(P_n)$.

Instead of working in $\mathbb{Q}^{\tau_n}$, to check X is in $conv(P_n)$ or not, I fold X on a two dimensional lattice formed by $k \in [4, n]$ and $e = (i, j) \in E_{n-1}$. The points $(k, (i, j))$, $j < k$ have weights $x_k(e) = x_{ijk} > 0$. Figure 5.1 depicts this folding.

I consider taking adjacent layers for solving a *forbidden arc transportation problem*, a bipartite flow problem, with certain arcs having zero capacity called forbidden arcs (see Definition 2.3). Depending on the feasibility of the problem, I decide to proceed further or not. Hence, any degree of non-determinism entering in the process is avoided.

I shall elaborate on the construction of the layered network. Prove the validity of the decisions made at each stage and explain the sub problems that are to be constructed and solved. This involved process requires careful understanding at each step. To have a complete proof of correctness, I will not leave anything for guessing. Sometimes the obvious is also written fully, for this reason. (I apologise for not being frugal. However, one may skip the proofs if it is clear or obvious from the definitions and the statement of the results presented.)

 Note *Recall that our aim is to express X as a convex combination of pedigrees in P_n or show evidence it cannot be done. For some X a certain pedigree(s) might appear in every convex combination possible. Therefore, identifying them and treating them exclusively becomes important. Also, because of the recursive nature of pedigrees, if we fail early, that is, show $X/k \notin conv(P_k)$ for some $k < n$, we stop. This is the essence of the approach.*

T. S. Arthanari, *Pedigree Polytopes*,
https://doi.org/10.1007/978-981-19-9952-9_5

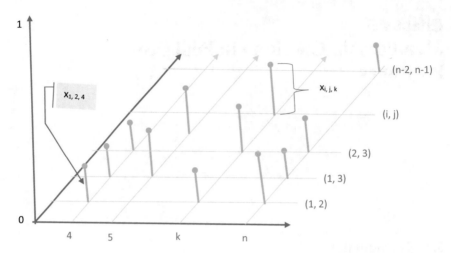

Fig. 5.1 Folding $X \in Q^{T_n}$ on a 2-dimensional lattice. Weights, x_{ijk}, for $(i, j) \in E_{k-1}$, and $4 \leq k \leq n$ are as depicted

5.2   Construction of the Layered Network

In this section, we define the layered network N_k and a set of pedigrees R_k with fixed positive weight μ_P for $P \in R_k$, with respect to a given $X \in P_{MI}(n)$ and for $k \in V_{n-1} \setminus V_3$.

We denote the node set of N_k by $\mathcal{V}(N_k)$ and the arc set by $\mathcal{A}(N_k)$. Let $v = [k : e]$ denote a node in the $(k - 3)$rd layer corresponding to an edge $e \in E_{k-1}$. Let $x(v) = x_k(e)$ for $v = [k : e]$.

5.2.1   Construction of the Network for $k = 4$

As shown in Sect. 4.4, $X \in conv(P_n)$, implies $X/n - 1 \in conv(P_{n-1})$. Therefore, if $X/5 \notin conv(P_5)$ we can conclude that $X \notin conv(P_n)$. $FAT_4(\mathbf{x_4})$ feasibility is both necessary and sufficient for $X/5 \in conv(P_5)$.

Let
$$V_{[r]} = \{v | v = [r + 3 : e], e \in E_{r+2}, x(v) > 0\}.$$

The node name $[r + 3 : e]$ alludes to the insertion decision corresponding to the stage r; that is, the edge e is used for insertion of $r + 3$. First we define the nodes in the network N_4.
$$\mathcal{V}(N_4) = V_{[1]} \bigcup V_{[2]}.$$

And

$$A(N_4) = \{(u, v) | u = [4 : e_\alpha] \in V_{[1]}, v = [5 : e_\beta] \in V_{[2]}, e_\alpha \in G(e_\beta)\}.$$

Let $x(w)$ be the capacity on a node $w \in V_{[r]}$, $r = 1, 2$. Capacity on an arc $(u, v) \in A(N_4)$ is $x(u)$.

Note *The nodes in $V_{[1]}$, $V_{[2]}$ are same as the origin and destination nodes in $FAT_4(\mathbf{X_4})$ problem defined and studied in Chap. 4, for consistency with subsequent development of the layered network for other $k > 4$, we choose these notations, and refer to this problem as F_4 in later discussions.*

Given this network, we consider a flow feasibility problem of finding a nonnegative flow defined on the arcs that saturates all the node capacities and violates no arc capacity, that is, solve $FAT_4(\mathbf{x_4})$ problem. Thus, the construction of the layered network proceeds from solving a bipartite flow feasibility problem at stage 1, involving layer 1 and 2, namely $FAT_4(\mathbf{x_4})$ problem. If $FAT_4(\mathbf{x_4})$ is infeasible we conclude $X \notin conv(P_n)$, and the construction of the network stops. When $FAT_4(\mathbf{x_4})$ is feasible we use the *FFF* algorithm (or any such) described in Sect. 2.5.1, and identify the set of rigid arcs, $\mathcal{R}$. If there are any dummy arcs in $\mathcal{R}$ we delete them from the set of arcs, and update. For every rigid arc with positive frozen flow we replace the capacity of the arc by the frozen flow. And include the corresponding arc (pedigree, $P \in conv(P_5)$) in the set R_4 with μ_P, the frozen flow as the weight for P. We reduce the node capacities of the head and tail nodes of the rigid arc P by μ_P, for each $P \in R_4$, and delete nodes with zero capacity, along with the arcs incident with them. Consider the subnetwork obtained as N_4. Let $\bar{x}(v)$ be the updated capacities of node v in the network. Now we say (N_4, R_4, μ) is *well-defined*. Observe that either the node set in N_4 is $\emptyset$ or $R_4 = \emptyset$, but not both.

Example 5.1 Consider $X \in P_{MI}(5)$ given by

$$\mathbf{x_4} = (0, 3/4, 1/4);$$
$$\mathbf{x_5} = (1/2, 0, 0, 0, 0, 1/2);$$

It can be verified that $FAT_4(\mathbf{x_4})$ is feasible and f given by $f_{([4:1,3],[5:1,2])} = 1/4$, $f_{([4:2,3],[5:1,2])} = 1/4$, and $f_{([4:1,3],[5:3,4])} = 1/2$ does it. And so, we conclude $X \in conv(P_5)$. But no arc is rigid in $FAT_4(\mathbf{x_4})$, is obvious. Therefore, $(N_4 = F_4, R_4 = \emptyset, \mu = 0)$. However, X' given by

$$\mathbf{x'_4} = (0, 3/4, 1/4);$$
$$\mathbf{x'_5} = (0, 1/4, 0, 0, 1/4, 1/2);$$

is such that, $X' \in P_{MI}(5)$ but $FAT_4(\mathbf{x_4})$ is not feasible. So $X' \notin conv(P_5)$.

Example 5.2 Consider Example 4.2, we have $FAT_4(\mathbf{x_4})$ feasible but all arcs are rigid and so $(N_4 = \emptyset, R_4 = \{((1, 3)(1, 4)), ((1, 3)(3, 4)), ((2, 3)(1, 3)), ((2, 3)(3, 4)), ((2, 3)(2, 4))\}, \mu = (\frac{1}{6}, \frac{1}{6}, \frac{1}{6}, \frac{1}{6}, \frac{1}{3})$.

5.2.2 Overview of the Membership Checking in Pedigree Polytope

If (N_{k-1}, R_{k-1}, μ) is well defined and $k < n$ we proceed further. Next, consider a bipartite flow feasibility problem for stage $k - 3$, called F_k. If F_k is infeasible we conclude $X \notin conv(P_n)$, and the construction of the network stops. Otherwise, we proceed to construct (N_k, R_k, μ). Next, we seek a feasible solution to a multi-commodity flow problem. If we have a feasible solution, we have evidence that $X/k + 1 \in conv(P_{k+1})$, and we say the network is well-defined. If (N_k, R_k, μ) is not well-defined, we conclude $X \notin conv(P_n)$, and stop. Otherwise proceed until we reach $k = n - 1$, constructing (N_k, R_k, μ), and concluding $X \in conv(P_n)$ or not.

Fig. 5.2 Flowchart—
Membership Checking in
$conv(P_n)$

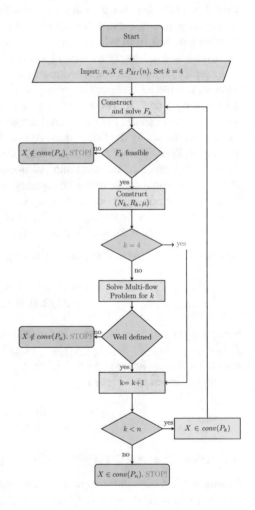

This construction process is explained in Fig. 5.2 and the following discussion elaborates on these steps and proves the necessary results to ensure that the conclusion made on the status of X's membership in the pedigree polytope is valid.

5.3 Construction of the Layered Network for k > 4

Given (N_{k-1}, R_{k-1}, μ) is well-defined, we proceed to construct (N_k, R_k, μ) recursively. Firstly, we define the node set,

$$\mathcal{V}(N_k) = \mathcal{V}(N_{k-1}) \bigcup R_{k-1} \bigcup V_{[k-2]}. \tag{5.1}$$

Definition 5.1 (*Link*) Let $l \in V_{n-2} \setminus V_3$ and $(e, e') \in E_{l-1} \times E_l$. Given $X \in P_{MI}(n)$, we say $([l : e], [l + 1 : e'])$ is a link in case 1. $x_l(e) > 0$ and $x_{l+1}(e') > 0$ and 2. either $e' \in E_l \setminus E_{l-1}$ and $e \in G(e')$, or $e, e' \in E_{l-1}$ and $e \neq e'$. If l is fixed or clear from the context, (e, e') is used as a link, for brevity of notation. A link can be used to extend a pedigree from P_l to a pedigree in P_{l+1}. ♣

In $\mathcal{V}(N_k)$ any pedigree, P in R_{k-1} is *shrunk* to become a new node with capacity μ_P. And we have the new sinks corresponding to $V_{[k-2]}$. The new origin node P introduced corresponding to a rigid pedigree ensures that the total weight on the extensions of P is exactly equals μ_P. We shall have an example to illustrate the importance of these additional origins. Now consider the links between layers $k - 3$ and $k - 2$. Any link, L can give raise to an arc in the network N_k depending on the solution to a max flow problem defined on a sub network (called restricted network) derived from N_{k-1} and the link L. The motivation for the restricted network stems from the fact, any link $L = ([k : (r, s)], [k + 1 : (i, j)])$ to extend a pedigree $P \in P_k$, we require (1) a generator of (r, s) as well as that of (i, j) must be present in P; (2) (r, s) and (i, j) should not be present in $P/k - 1$ and (3) P must end in (r, s). So we need not consider paths in N_{k-1} that:

- are not ending in $[k : (r, s)]$.
- are having $[q : (r, s)]$ or $[q : (i, j)]$ for some $q < k$.
- are not having a generator node of (r, s) or a generator node of (i, j).

The restricted network $N_{k-1}(L)$ defined for a link L ensures that it does not have such paths in it, by deleting a set of nodes from N_{k-1}.

We define the restricted network for a link L, which is induced by the deletion of a subset of nodes from $\mathcal{V}(N_{k-1})$.

Definition 5.2 (*Restricted Network*) Given $k \in V_{n-1} \setminus V_4$, a link $L = ([k : e_\alpha], [k + 1 : e_\beta])$, with $e_\alpha = (r, s) \in E_{k-1}$ and $e_\beta = (i, j) \in E_k$. $N_{k-1}(L)$ is the sub network induced by the subset of nodes $V(N_{k-1}) \setminus \mathbf{D}$, where $\mathbf{D}$, the set of deleted nodes, is constructed as follows:

Let $\mathbf{D} = \emptyset$.

Deletion Rules:

(a) Include $[l : e_\beta]$ in $\mathbf{D}$, for $max(4, j) < l < k$.

(b) Include $[l : e_\alpha]$ in $\mathbf{D}$, for $max(4, s) < l < k$.

(c) Include $[j : e], e \notin G(e_\beta)$ in $\mathbf{D}$, if $e_\beta \in E_k \setminus E_3$; otherwise include $[4 : e_\beta]$ in $\mathbf{D}$.

(d) Include $[s : e], e \notin G(e_\alpha)$ in $\mathbf{D}$, if $e_\alpha \in E_{k-1} \setminus E_3$; otherwise include $[4 : e_\alpha]$ in $\mathbf{D}$.

(e) Include all nodes $V_{[k-3]} \setminus \{[k : e_\alpha]\}$ in $\mathbf{D}$.

(f) If any undeleted node $[l : e], l > 4$ is such that, all its generators have been deleted, include that node in $\mathbf{D}$. (Notice that any undeleted nodes $[4 : e]$ is an origin node and the rule is not applied on such a node.)

Repeat this step until no more nodes are deleted. Set $V(N_{k-1}(L)) = V(N_{k-1}) \setminus \mathbf{D}$.

(g) If a node from $\mathbf{D}$ appears in any of the rigid pedigree paths (corresponding to any l), such pedigrees are deleted. This means the corresponding (shrunk) nodes are deleted from $V(N_{k-1}(L))$.

The subnetwork induced by $V(N_{k-1}(L))$ is called the *Restricted Network* $N_{k-1}(L)$.   ♣

Recall that a vertex-induced subnetwork is a subset of the nodes of a network together with any edge whose both endpoints are in this subset.

Remark A few observations that are easy to make are listed below:

1 Deletion rule [a] ([b]) ensures that the node $[l : e_\beta]$ ($[l : e_\alpha]$) does not appear earlier in a path from source(s) in layer 1 to $(k - 3)$rd layer. Deletion rule [c] ([d]) ensures that the edge(s) not in the generator of the edge e_β (e_α) are deleted from $(j - 3)$rd$((s - 3)$rd) layer. Deletion rule [e] ensures that the only sink in $(k - 3)$rd layer is $[k : e_\alpha]$. Rule [f] defines the nodes in the restricted network, ensuring each node in the restricted network has at least one generator. Lastly, [g] ensures that in any stage l, a rigid pedigree is deleted from R_l, if the corresponding pedigree path contains a deleted node. So the corresponding (shrunken) node is deleted from the restricted network.

2 The deletion of a node can be equivalently interpreted as imposing an upper bound of zero on the flow through a node for a given link (treated as a commodity). This interpretation is useful in considering a multicommodity flow through the network N_k discussed in the later sections.

3 Construction of the restricted network for any link can be done by applying deletion rules a through e building the set $\mathbf{D}$. Then we need only to check whether any node in $\mathcal{V}(N_{k-1}) \setminus \mathbf{D}$ is deleted using deletion rule f. We check the undeleted nodes layer by layer in $V_{[l-3]}, 5 \leq l \leq k$. If any node is deleted as per rule f, it can only affect the generator availability status of nodes in the subsequent layers and not in the current layer or the layers already checked applying rule f. So while applying rule f whenever $\mathbf{D}$ is updated, it is sufficient to check rule f for the remaining undeleted nodes.

Therefore, to obtain finally $\mathcal{V}(N_{k-1})(L)$, deletion rule f is repeated at most $|\mathcal{V}(N_{k-1})|$ times.

4 Please note we do not permanently delete any node in $\mathcal{V}(N_{k-1})$. This is a ploy for computing the capacity for a link L.

5.3.1 Capacity Along a Link

To determine the capacity along a link L, we consider the problem of finding the maximal flow in $N_{k-1}(L)$ satisfying non-negativity of flow, flow conservation at nodes, and capacity restrictions on the available nodes and arcs.

The only sink in the network is $[k : e_\alpha]$ and the sources are the undeleted nodes in $V_{[1]}$ and undeleted pedigrees in $R_l, 4 \leq l \leq k - 2$. Let $C(L)$ be the value of the maximal flow in the restricted network $N_{k-1}(L)$. We find $C(L)$ for each link L.

Besides, if the path that brings this flow $C(L)$ for any L is unique, we save the corresponding path, named $P_{unique}(L)$, against L. By abuse of notation, we also use $P_{unique}(L)$ to denote the corresponding ordered set of edges along the path.

Remark Notice that the complexity of computing $C(L)$ (and saving the unique path if need be) for any L is the same as the complexity of finding a maximal flow in the network $N_{k-1}(L)$. And is low order strongly polynomial in the number of nodes and edges in the network [108].

Now we are in a position to define the *FAT* problem, called F_k.

Definition 5.3 Consider a forbidden arc transportation problem with

$O--$	Origins]	$: u = [k : e_\alpha] \in V_{[k-3]}$ and $P \in R_{k-1}$
$a--$	Availability]	$: \bar{x}(u)$ for u and μ_P for P
$D--$	Sinks]	$: v = [k+1 : e_\beta] \in V_{[k-2]}$
$b--$	Demand]	$: x(v)$ for v
$\mathcal{A}--$	Arcs]	$: \{(u, v)$ such that, either $L = (e_\alpha, e_\beta)$ is a link and $C(L) > 0$
	]	$:$ or $u = P$ and $v = [k+1 : e_\beta]$ and (P, e_β) is a pedigree,
	]	$:$ that is, $(P, e_\beta) \in P_{k+1}\}$
$C--$	Capacity]	$: C_{u,v} = C(L)$ or μ_P depending on u.

♣

If F_k is feasible we apply the *Frozen Flow Finding* algorithm and identify the set $\mathcal{R}$ and the dummy subset of arcs in that. Update $\mathcal{A}$ by deleting the dummy arcs. Update the capacity of any rigid arc, $C(L)$, to the frozen flow. Recall that the frozen flow is a positive constant, in every feasible solution to F_k.

Lemma 5.1 *If $L = (u, v)$ is rigid and the path is unique for L and $(P_{unique}(L),$ $[k + 1 : e_\beta])$ does not correspond to a pedigree then we can conclude that $X/k + 1 \notin conv(P_{k+1})$.* ♡

Proof The condition implies that it is impossible to write $X/k + 1$ as a convex combination of pedigrees in P_{k+1} alone as $(P_{unique}(L), e_\beta)$ receives the positive weight, $C_L > 0$. Hence the result.

However we shall see later if F_k is feasible, such an unique non-pedigree path is impossible corresponding to a rigid link (Theorem 5.2).

Definition 5.4 (*Pedigree path*) Consider the network N_l and a path P in it. If P corresponds to a pedigree $X^r \in P_{l+1}$, where X^r is the characteristic vector of $P^r = (e_4^r, \ldots e_{l+1}^r)$, we call it a pedigree path. ♣

Thus, with respect to (N_4, R_4, μ), pedigrees in R_4 correspond to rigid pedigree paths. They receive fixed positive weights in all convex combinations of pedigrees expressing $X/5$. And in general $P \in R_k$ correspond to a rigid pedigree path and it receives fixed weight in all convex combinations of pedigrees in P_{k+1} expressing $X/k + 1$.

5.3.2   Completing the Construction of the Layered Network

In Sect. 5.4 we shall show that if F_k is not feasible $X/k + 1 \notin conv(P_{k+1})$. Then as per the flowchart shown in Sect. 5.2.2 we do not proceed further. Assuming F_k is feasible, for now, we are in a position to complete the construction of the network (N_k, R_k, μ). Construction of R_k is driven by the rigid arcs in F_k. If the rigid flow is from a unique path, then that contributes a rigid pedigree to R_k. Similarly, if the rigid flow is from an origin in F_k that is a shrunken node, $P \in R_{k-1}$ then that contributes a rigid pedigree to R_k. This construction is formally expressed below: First, we construct R_k, (the set of rigid pedigrees), and μ, (their capacities). Let $L = (u = [k : e_\alpha], v = [k + 1 : e_\beta])$.

Case 1: For any rigid arc L from F_k, with a unique path, $P_{unique}(L)$
The unique path, $P_{unique}(L)$ for any link L is necessarily a pedigree path, otherwise we would not proceed, as we know $X/k + 1 \notin conv(P_{k+1})$ from Lemma 5.1. For

any such rigid arc L from F_k, we include the extended pedigree $(P_{unique}(L), e_\beta)$ to R_k^1.

$$R_k^1 = \{ (P_L, e_\beta) : L \text{ is rigid } \& P_L \text{ is unique for } L. \} \tag{5.2}$$

For each $P \in R_k^1$ set $\mu_P = C(L)$, where L is the defining rigid arc. Update $\mathcal{A}$, (the set of arcs in F_k, see Definition 5.3) by deleting the defining rigid arcs. Update the node capacity of any node $u \in \mathcal{V}(N_k)$ that appears in some rigid pedigree paths, that is, $path(P)$, $P \in R_k$, by reducing the node capacity of u by the rigid flows (μ_P's) along these paths. That is,

$$\text{updated capacity}(u) = \text{capacity}(u) - \sum_{P \in R_k^1 | u \in path(P)} \mu_P, u \in \mathcal{V}(N_k). \tag{5.3}$$

Similarly, it is important that we update the *arc capacities* along the unique pedigree path, $P_{unique}(L)$ for a rigid L.

Case 2: For any rigid arc $L = (P', v = e_\beta)$
For any rigid arc connecting a pedigree $P' \in R_{k-1}$, we include the extended pedigree (P', e_β) to R_k^2.

$$R_k^2 = \{(P', e_\beta) : P' \in R_{k-1} \text{ and } L \text{ is rigid.}\} \tag{5.4}$$

For each rigid pedigree, $P \in R_k^2$ (as per Eq. 5.4), set $\mu_P = C(L)$, where L is the defining rigid arc. Update $\mathcal{A}$, by deleting the defining rigid arcs. Update the node capacity of any node $u = P' \in R_{k-1}$ that appears in some rigid pedigree, $P \in R_k^2$, by reducing the node capacity of u by the rigid flows (μ_P's). That is,

$$\text{updated capacity}(u = P') = \text{capacity}(u) - \sum_{P \in R_k^2 | P \text{ is extended from } P'} \mu_P, P' \in R_{k-1}. \tag{5.5}$$

Now set,

$$R_k = R_k^1 \cup R_k^2 \tag{5.6}$$

Remark We can assume without loss of generality that we have distinct pedigrees in R_k. Otherwise, we can merge the weights of multiple occurrences of a pedigree together and ensure that the pedigrees are all distinct.

Next, we update the node set. In Eq. 5.1 we defined the node set for N_k as,

$$\mathcal{V}(N_k) = \mathcal{V}(N_{k-1}) \bigcup R_{k-1} \bigcup V_{[k-2]}.$$

If any node (including shrunken nodes for rigid pedigrees in any stage) in the network has zero capacity it is deleted from the node set. Let $\mathcal{D}_0$ be the set of such deleted nodes, with an updated capacity equal to zero. Let $\mathcal{A}_{\mathcal{D}_0}$ denote set of arcs incident on any node in $\mathcal{D}_0$. We delete these arcs from the set of arcs in N_{k-1}. That is, we consider the sub network induced by the updated $\mathcal{V}(N_k)$.

$$\mathcal{V}(N_k) := \text{(is updated as) } \mathcal{V}(N_k) \backslash \mathcal{D}_0. \tag{5.7}$$

and

$$\mathcal{A}(N_k) := \text{(is updated as) } \mathcal{A}(N_{k-1}) \backslash \mathcal{A}_{\mathcal{D}_0} \tag{5.8}$$

One last thing to do is to include remaining arcs, $\mathcal{A}$ of F_k to the network. Thus,

$$\mathcal{A}(N_k) = \mathcal{A}(N_k) \bigcup \mathcal{A}. \tag{5.9}$$

Thus, we have constructed recursively the network (N_k, R_k, μ), where N_k is given by Eqs. 5.7 and 5.9, and capacities of nodes from (5.3) and (5.5). R_k is obtained as given by Eq. 5.6, with μ_P as capacity for $P \in R_k$.

Definition 5.5 Let $\bar{x}(u)$ denote the updated capacity for $u = [l : e] \in N_k$, and let $\bar{\mu}_{P'}$ give the updated capacity for $u = P' \in \bigcup_{l=4}^{k-1} R_l$. Let in general $c(u)$ denote the capacity of any node in N_k irrespective of $u = [l : e]$ or $P' \in R_l$.   ♣

It is easy to show the following lemma.

Lemma 5.2 *For any $[l : e] \in N_k, 4 \leq l \leq k + 1$, we have*

$$x([l : e]) = \bar{x}([l : e]) + [\sum_{P' \in \bigcup_{l=4}^{k-1} R_l \,|\, [l:e] \in path(P')} \bar{\mu}_{P'}] + \sum_{P \in R_k \,|\, [l:e] \in path(P)} \mu_P.$$

Let $\hat{x}([l : e]) = \sum_{P' \in \bigcup_{l=4}^{k-1} R_l \,|\, [l:e] \in path(P')} \bar{\mu}_{P'} + \sum_{P \in R_k \,|\, [l:e] \in path(P)} \mu_P.$ Then equivalently,

$$x([l : e]) = \bar{x}([l : e]) + \hat{x}([l : e]).$$

                                                                                     ♡

We understand $\hat{X}$ as the vector in R^{τ_k}, whose typical element is $\hat{x}([l : e])$. The total weight of rigid pedigrees that have edge e in position $l - 3$ is given by $\hat{x}([l : e])$. Equivalently, $\hat{x}([l : e])$ is the total capacity earmarked for rigid pedigree paths that have $[l : e]$ in them.

A schematic diagram of the network N_k and R_k is shown in Fig. 5.3.

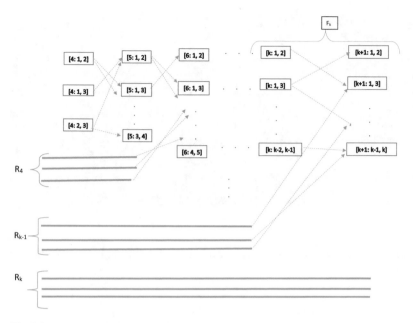

Fig. 5.3 Schematic diagram of the network, N_k, R_k

5.4 A Sufficient Condition for Non-membership

The importance of studying the problem F_k is discussed in this section. For this purpose, we define another *FAT* problem for which an obvious feasible flow, called the *instant flow*, is available, and this flow can be very useful in proving the feasibility of F_k. This problem is used only in the proofs of certain results and not required to be solved, like F_k, for a given instance.

Definition 5.6 Given $X/k + 1 \in conv(P_{k+1})$, $\lambda \in \Lambda_{k+1}(X)$ we define a *FAT* problem obtained from $I(\lambda)$ for a given $l \in \{4, \ldots, k\}$ called, $INST(\lambda, l)$ as follows:

Partition $I(\lambda)$ in two different ways according to $\mathbf{x}_l^r$, $\mathbf{x}_{l+1}^r$, resulting in two partitions S_O and S_D. We have

$$S_O^q = \{r \in I(\lambda) | x_l^r(e_q) = 1\}, e_q \in E_{l-1} \text{ and } x_l(e_q) > 0$$

and

$$S_D^s = \{r \in I(\lambda) | x_{l+1}^r(e_s) = 1\}, e_s \in E_l \text{ and } x_{l+1}(e_s) > 0.$$

Let $|y|_+$ denote the number of positive coordinates of any vector y. Let $n_O = |\mathbf{x}_l|_+$ and $n_D = |\mathbf{x}_{l+1}|_+$. Let $a_q = \sum_{r \in S_O^q} \lambda_r = x_l(e_q), q = 1, \ldots, n_O$. Let $b_s = \sum_{r \in S_D^s} \lambda_r = x_{l+1}(e_s), s = 1, \ldots, n_D$. Let the set of forbidden arcs, F, be given by

$$F = \{(q, s) | S_O^q \cap S_D^s = \emptyset\}.$$

The problem with an origin for each q with availability a_q, a sink for each s with demand b_s and the forbidden arcs given by F is called the $INST(\lambda, l)$ problem.[a] ♣

---

[a] Construction of $INST(\lambda, l)$ problem is same as that done in the proof of Theorem 4.2.

Remark

1. From Lemma 2.1 we know that such a problem is feasible and a feasible flow is given by

$$f_{q,s} = \sum_{r \in S_O^q \cap S_D^s} \lambda_r.$$

   We call such an f the *instant flow* for the $INST(\lambda, l)$ problem.
2. The sinks and the demands in the above problem are same as that of F_l.
3. But in the problem F_l, for an origin $[l : e_\alpha]$ we have availability as the updated $x_l(e_\alpha)$, that is, $\bar{x}([l : (e_\alpha)]$, and for $P \in R_{l-1}$ we have $\bar{\mu}_P$ as the availability. But there are no capacity restrictions on the arcs of the $INST(\lambda, l)$ problem.

We shall show that the instant flow for the $INST(\lambda, l)$ problem is indeed feasible for the problem F_l.

Definition 5.7 Given $\lambda \in \Lambda_{k+1}$, and l, for $r \in I(\lambda)$, let L_l^r denote (e, e') such that $x_l^r(e) = x_{l+1}^r(e') = 1$. That is, L_l^r is the $(l-3)$rd and $(l-2)$nd elements of the pedigree P^r (given by X^r). ♣

Lemma 5.3 Given $\lambda \in \Lambda_{k+1}$, and l, if for $r \in I(\lambda)$, either $path(X^r/l)$ is available in $N_{l-1}(L_l^r)$, or corresponds to a pedigree in R_{l-1}, then the instant flow f for the $INST(\lambda, l)$ problem is feasible for F_l. ♡

Proof Given the instant flow for $INST(\lambda, l)$, consider the flow $f_{q,s}$ along an arc $L = ([l : e_q], [l + 1 : e_s])$. We shall show that this flow can be split between different arcs connecting $[l + 1 : e_s]$ and origins in F_l, that have $[l : e_q]$ in common, without violating any arc capacities.

Let $\mathbb{S} = \{r \in I(\lambda) | X^r / l$ corresponds to a $P \in R_{l-1}\}$.
First we consider the origins in F_l not corresponding to $P \in R_{l-1}$. Now consider,

$$r \in \{S_O^q \setminus \mathbb{S}\} \cap S_D^s \neq \emptyset, \text{ for some } q, s.$$

For all these r, $L_l^r = (e_q, e_s)$, and is available in the $INST(\lambda, l)$ problem. For any such r, it is given that $path(X^r / l)$ is available in $N_{l-1}(L_l^r)$. Along this $path(X^r / l)$ we can have a flow of λ_r into $[l : e_q]$. So we are ensured that the maximum flow, $C(L)$, in $N_{l-1}(L)$ is positive for $L = L_l^r$. According to the construction of problem F_l we have an arc $([l : e_q], [l+1 : e_s])$ with capacity $C(L)$. Now the definition of the instant flow f and the maximality of $C(L)$ imply that f does not violate any of the capacity restrictions. Since a flow of at least $\sum_{r \in \{S_O^q \setminus \mathbb{S}\}} \lambda_r$ can reach $[l : e_q]$. So the maximum flow, $C(L)$ in $N_{l-1}(L)$ should be at least this.

Next, we consider the origins in F_l corresponding to $P \in R_{l-1}$. We partition R_{l-1} with respect to e_q. Let $R_{l-1,q} = \{P \in R_{l-1} |$ the last component of P is $e_q\}$. Some of these sets might be empty. For $P \in R_{l-1}$, let $\mathbb{S}_P = \{r \in \mathbb{S} | X^r / l = P\}$. A flow of $\sum_{P \in R_{l-1,q}} \sum_{r \in \mathbb{S}_P} \lambda_r$ can reach $[l : e_q]$.

Necessarily, $X/l \in conv(P_l)$ as $X/k + 1 \in conv(P_{k+1})$. We know that while expressing X/l as a convex combination of pedigrees in P_l, in any such combination, each $P \in R_{l-1}$ receives the weight $\bar{\mu}_P$ being a rigid pedigree. Hence, $\sum_{r \in \mathbb{S}_P} \lambda_r = \bar{\mu}_P$ for any $P \in R_{l-1,q}$ and any e_q, where $\bar{\mu}_P$ is the residual capacity of any $P \in R_{l-1}$.

Now, in F_l, we have an arc $(P, [l+1 : e_s])$ for a $P \in R_{l-1,q}$ if (P, e_s) is an extension of P. The flow along any such arc in F_l will be equal to $\sum_{r \in \mathbb{S}_P \cap S_D^s} \lambda_r$.

Thus the flow along (e_q, e_s) in $INST(\lambda, l)$, $f_{q,s}$, is split between the different origins in F_l that have e_q in common. Thus we have shown that the instant flow of $INST(\lambda, l)$ is feasible for F_l. Hence the lemma.

Note *Note that in this proof, firstly, we have partitioned $I(\lambda)$ into $\mathbb{S}$ and the rest. Then while considering shrunk nodes corresponding to $P \in R_{l-1}$, we partition R_{l-1} into $R_{l-1,q}, q \ni P \in R_{l-1}$ and ends in e_q. We also partition $\mathbb{S}$ into $\mathbb{S}_P$ for $P \in R_{l-1}$. Thus,*

$$\bigcup_q \bigcup_{P \in R_{l-1,q}} \mathbb{S}_P = \mathbb{S}.$$

This way every r in $I(\lambda)$ is accounted for.

Next, we address the question: Are the conditions stated by Lemma 5.3 true for all pedigrees active for $X/k + 1 \in conv(P_{k+1})$?

Lemma 5.4 *Every $P^* = (e_4^*, \ldots, e_{k+1}^*)$ active for $X/k + 1$, is such that, either [a] $P^*/5$ is available in R_4 or [b] $path(P^*/5)$ is available in $N_4(L_5^*)$, where $L_5^* = (e_5^*, e_6^*)$.* ♡

Proof $Path(P^*/5)$ is given by $[4 : e_4^*] \rightarrow [5 : e_5^*]$. We have $[5 : e_5^*]$ in $N_4(L_5^*)$ as the lone sink. If the result is not true, we have either

Case 1: $[4 : e_4^*]$ is not a node in $N_4(L_5^*)$,
For this case to happen, there exists a deletion rule among the rules given in Definition 5.2 that deleted $[4 : e_4^*]$ from $\mathcal{V}(N_4)$. Rules (a), (b), (e) and (f) are not applicable as they delete a node $[l : e]$ with $l > 4$. We shall see that rule (c) or (d) does not delete $[4 : e_4^*]$.

Suppose $e_6^* = (i, j) \in E_5 \setminus E_3$ and $j = 5$. Then rule (c) deletes a node $[5 : e]$. Suppose $e_6^* = (i, 4)$ for some $1 \leq i < 4$, then rule (c) deletes $[4 : e], e \notin G(e_6^*)$. So if $[4 : e_4^*]$ is one such node then e_4^* is not a generator of e_6^*. And so $X^*/6$ can not be in P_6, leading to a contradiction. This implies $e_6^* \in E_3$. Then $[4 : e_6^*]$ is deleted. So rule (c) does not delete $[4 : e_4^*]$.

Similarly, we can check that rule (d) does not delete $[4 : e_4^*]$. In other words, Case 1 is not possible.

Case 2: $[4 : e_4^*]$ exists but $Path(P^*/5)$ is not in $N_4(L_5^*)$.
Suppose $([4 : e_4^*], [5 : e_5^*])$ is not an arc in $N_4(L_5^*)$. P^* being a pedigree, implies that $e_4^* \in G(e_5^*)$. So this arc exists in F_4 with capacity $x_4(e_4^*) > 0$. Necessarily F_4 is feasible as $X/k + 1$ is assumed to be in $conv(P_{k+1})$. Since X^* is active for $X/k + 1$, we have a $r \in I(\lambda)$ such that $X^* = X^r$, for some $\lambda \in \Lambda_{k+1}(X)$. Consider the problem $INST(\lambda, 4)$. So we have the instant flow f that is feasible for F_4. Thus, we are eligible to apply FFF algorithm to find dummy arcs in F_4, towards constructing N_4.

The flow along $([4 : e_4^*], [5 : e_5^*])$ is positive as the corresponding set $S_O^q \cap S_D^s \neq \emptyset$, (see Definition 5.6) has at least r corresponding to X^* in it. This ensures that the arc $([4 : e_4^*], [5 : e_5^*])$ is not a dummy. And so it results in [b]. Hence Case 2 is not possible.

However, if $([4 : e_4^*], [5 : e_5^*])$ is rigid, in which case, $P^*/5$ is included in R_4 so, we have proved that it results in [a]. This completes the proof of the lemma.

Lemma 5.4 forms the basis to prove Lemma 5.5 that is crucial in showing that infeasibility of F_k implies that $X/k + 1 \notin conv(P_{k+1})$.

Lemma 5.5 (Existence of Pedigree Paths) *Every X^* active for $X/k + 1$, is such that, either [a] P^*/l is in R_{l-1}, or [b] $path(P^*/l)$ is available in $N_{l-1}(L_l^*)$, for $5 \leq l \leq k$, where L_l^* denotes (e, e') such that $x_l^*(e) = x_{l+1}^*(e') = 1$.* ♡

Proof If any X^* active for $X/k + 1 \in conv(P_{k+1})$ corresponds to (P, e_β) for some $P \in R_{k-1}$, then there is nothing to prove. So, assume this is not the case.

[Proof by induction on l] From Lemma 5.4 we have the result for $l = 5$. Assume that the result is true for $l \leq k - 1$. We shall show that the result is true for $l = k$. Now consider $N_{k-1}(L_k^*)$. Suppose $path(X^*/k)$, that is,

$$[4 : e_4^*] \rightarrow \ldots [k - 1 : e_{k-1}^*] \rightarrow [k : e_k^*]$$

is not available in $N_{k-1}(L_k^*)$ or is not a rigid pedigree in R_{k-1}. However, $[k : e_k^*]$ is available in $N_{k-1}(L_k^*)$ as it is the lone sink.

By the induction hypothesis we have $path(X^*/k - 1)$ available either [1] in $N_{k-2}(L_{k-1}^*)$ or [2] corresponds to a rigid pedigree in R_{k-2} ending in $[k - 1 : e_{k-1}^*]$.

If [2] is true, then $(path(X^*/k - 1), [k : e_k^*])$ is an eligible arc in F_{k-1}, as (X^*/k) is a pedigree. If this arc is not rigid, then the corresponding pedigree is available in $N_{k-1}(L_k^*)$. If the arc is rigid, and the frozen flow is 0, that is, it a dummy arc, then this pedigree, X^*/k is not appearing in any convex combination of pedigrees expressing X/k. So $X^*/k + 1$ an extension of X^*/k, can not be active for $X/k + 1$. Contradiction. Therefore, if the arc is rigid, it must be in R_{k-1}. However, we have assumed this is not the case at the very beginning of the proof.

In that case, our assumption implies that the arc $W = ([k - 1 : e_{k-1}^*], [k : e_k^*])$ does not exist in $N_{k-1}(L_k^*)$.

This leads to two cases to consider.

Case 1: W does not exist in N_{k-1}.
We shall show that this case is not possible. The existence of $path(X^*/k - 1)$ in $N_{k-2}(L_{k-1}^*)$ implies that the maximum flow into the sink, $[k - 1 : e_{k-1}^*]$ is positive. So W is an arc in F_{k-1}. Since X^* is active for $X/k + 1$ we have for some $\lambda^* \in \Lambda_{k+1}(X)$ a $r_0 \in I(\lambda^*)$ such that $X^* = X^{r_0}$. Consider the $INST(\lambda^*, k - 1)$ problem for such a λ^*. Observe that every $r \in I(\lambda^*)$ is active for $X/k + 1$. So from the induction hypothesis $path(X^r/k - 1)$ is available in $N_{k-2}(L_{k-1}^r)$ for every $r \in I(\lambda^*)$. We have the condition of Lemma 5.3 met here. Consider the instant flow f that is feasible for $INST(\lambda^*, k - 1)$ problem. Lemma 5.3 asserts that f is indeed feasible for F_{k-1}.

But if W does not exist in N_{k-1} it might be because W has been subsequently declared dummy by the FFF algorithm. Since the flow along W as per f is at least equal to $\lambda_{r_0}^*$ corresponding to X^*, which agrees with $L_{k-1}^* = (e_{k-1}^*, e_k^*)$. But X^* is active for $X/k + 1$ means $\lambda_{r_0}^* > 0$. Therefore, $W = ([k - 1 : e_{k-1}^*], [k : e_k^*])$ can not be declared dummy by FFF algorithm. Thus, this possibility is ruled out. However, if W is rigid and a unique path P ends in $[k - 1 : e_{k-1}^*]$ then the pedigree corresponding to the extended path $(P, [k : e_k^*])$ is included in R_{k-1}. That proves [a].

Case 2: W does not exist in $N_{k-1}(L_k^*)$.
Observe that this implies that as a consequence of the deletion rules, W has been deleted (recall Definition 5.2). Let $\mathbf{D}$ be the set of nodes deleted. The sub network induced by the nodes $\mathcal{V}(N_{k-1}) \setminus \mathbf{D}$, is $N_{k-1}(L_k^*)$ by definition. Check that nodes $[k - 1 : e_{k-1}^*]$ and $[k : e_k^*]$ are in $N_{k-1}(L_k^*)$. Note that every arc with both of its ends available in $N_{k-1}(L_k^*)$ is available in the network. Or Case 2 is impossible.

This completes the proof of the lemma.

Remark This is a significant result as it shows that our construction of the restricted network does not destroy any pedigree path corresponding to a pedigree that is active for $X/k + 1$. Furthermore, it aids us in proving the following theorem that gives a sufficient condition for nonmembership in the pedigree polytope.

Theorem 5.1 (Theorem on nonmembership) *Given $X \in P_{MI}(n)$, and for a $k \in V_{n-1} \setminus V_3$, if $X/k \in conv(P_k)$, then*

$$F_k \text{ infeasible implies } X/k + 1 \notin conv(P_{k+1}). \qquad \heartsuit$$

Proof Suppose $X/k + 1 \in conv(P_{k+1})$. Consider any $\lambda \in \Lambda_{k+1}(X)$. Then from Lemma 5.5 we have the path corresponding to X^r/l available either in $N_{l-1}(L_l^r)$ or $P^r/l \in R_{l-1}$ for each $r \in I(\lambda)$ and $5 \leq l \leq k$. Now conditions of Lemma 5.3 are met. And so, the instant flow in the $INST(\lambda, l)$ problem is feasible for F_l, for $5 \leq l \leq k$. We have a contradiction. Hence the theorem.

Remark 1. With this theorem, we have a procedure to check $X/k + 1 \notin conv(P_{k+1})$ by solving F_k.
2. As a corollary to this theorem, we have, given $X/k \in conv(P_k)$ implies for any $\lambda \in \Lambda_k(X)$ we have all the paths corresponding to $\{X^r | r \in I(\lambda)\}$ in N_{k-1}.

Next example shows that F_k feasibility is not sufficient for membership in a pedigree polytope.

Example 5.3 Consider X given by

$$\begin{aligned}
\mathbf{x}_4 &= (0, 3/4, 1/4); \\
\mathbf{x}_5 &= (1/2, 0, 0, 0, 0, 1/2); \\
\mathbf{x}_6 &= (0, 1/4, 1/4, 0, 1/4, 1/4, 0, 0, 0, 0).
\end{aligned}$$

It can be verified that $X \in P_{MI}(6)$, checking recursively. And F_4 is feasible, see f given by

$$f_{([4:1,3],[5:1,2])} = 1/4, \; f_{([4:2,3],[5:1,2])} = 1/4, \; f_{([4:1,3],[5:3,4])} = 1/2$$

is one such feasible solution.

Or $X/5 \in conv(P_5)$ and notice that no arc in F_4 is rigid, that is, $R_4 = \emptyset$. Thus, N_4 is well-defined. In Fig. 5.4a we have N_4.

Next, the restricted networks $N_4(L)$ for the links in $\{(1, 2), (3, 4)\} \times \{(1, 3), (2, 3), (2, 4), (3, 4)\}$ are shown in Fig. 5.5, where L is given as the arc connecting a node in $V_{[2]}$ with a node in $V_{[3]}$. F_5 is feasible, with f given by

$$f_{([5:1,2],[6:1,3])} = f_{([5:1,2],[6:3,4])} = 1/4, \; f_{([5:3,4],[6:2,3])} = f_{([5:3,4],[6:2,4])} = 1/4.$$

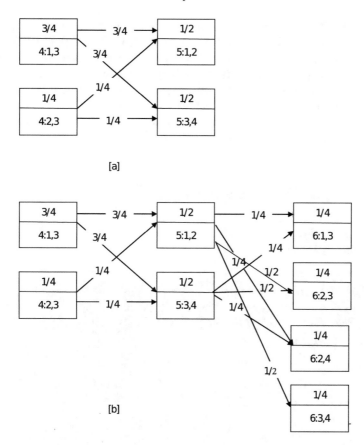

Fig. 5.4 Layered Networks—**a** N_4 and **b** N_5 for Example 5.3

Restricted Networks			
$N_4(L)$	L	C(L)	Unique Path?
[4:2,3] -> [5:1,2]	[5:1,2] -> [6:1,3]	1/4	Yes
[4:2,3] -> [5:3,4]	[5:3,4] -> [6:1,3]	1/4	Yes
[4:1,3] -> [5:1,2]	[5:1,2] -> [6:2,3]	1/2	Yes
[4:1,3] -> [5:3,4]	[5:3,4] -> [6:2,3]	1/2	Yes
[4:2,3] -> [5:1,2]	[5:1,2] -> [6:2,4]	1/4	Yes
[4:2,3] -> [5:3,4]	[5:3,4] -> [6:2,4]	1/4	Yes
[4:1,3] -> [5:1,2]	[5:1,2] -> [6:3,4]	1/2	No
[4:2,3] -> [5:1,2]			

Fig. 5.5 Restricted Networks $N_4(L)$ for Example 5.3

In Fig. 5.4b we have the layered networks for $k = 5$ and the capacity of an arc is shown along the arc. Note that both R_4 and R_5 are empty. We need to check whether N_5 is well-defined? Or Is $X/6$ a member of $conv(P_6)$?

Suppose $X/6$ is in $conv(P_6)$. Consider any $\lambda \in \Lambda_6(X)$, then λ necessarily assigns on the pedigrees corresponding to $([4:2,3],[5:1,2]),[6:1,3])$ and $([4:2,3],[5:3,4]),[6:1,3])$ a total weight of $1/4$ to saturate the capacity of $[6:1,3]$. This implies the capacity of $[4:2,3]$ is saturated as well. Similarly, there are pedigrees X^s, $s \in I(\lambda)$ such that $x_6^s(2,4) = 1$. The total weight for these pedigrees is $x_6(2,4) = 1/4$. But these pedigrees should have $x_4^s(2,3) = 1$ as $(2,3)$ is the only generator for $(2,4)$. However, this is not possible as the node capacity of $[4:2,3]$ is already saturated. Therefore $X \notin conv(P_6)$.

5.4.1 Pedigree Packability of Arc Flows

Suppose (N_{k-1}, R_{k-1}, μ) well-defined, so we have evidence for $X/k \in conv(P_k)$. Assume that F_k is feasible.

Definition 5.8 Consider any feasible flow, f in $N_{l-1}(L)$ for a link $L = (e_\alpha, e_\beta) \in E_{l-1} \times E_l$. Let v_f be the value of the flow f, that reaches the sink in $N_{l-1}(L)$. We say v_f is *pedigree packable* in case there exists a subset, $P(L) \subset P_l$ such that

(a) $\lambda_r (\geq 0)$ is the flow along $path(X^r)$ for $X^r \in P(L)$,
(b) $e_l^r = e_\alpha$, $X^r \in P(L)$,
(c) $\sum_{r \ni x^r(v)=1} \lambda_r \leq c(v)$, $v \in \mathcal{V}(N_{l-1}(L))$, and
(d) $\sum_{X^r \in P(L)} \lambda_r = v_f$.

where, $c(v)$ is the capacity for node v, and $c(v) = \bar{x}(v)$ if $v \in N_l$ or $c(v) = \bar{\mu}(v)$ if $v \in R_l$ for any l.
     We refer to $P(L)$ as a *pedigree pack* of v_f.   ♣

Definition 5.9 (*Extension Operation*) Given a pedigree pack corresponding to a flow f in $N_{l-1}(L)$ for a link $L = (e_\alpha, e_\beta)$ with $v_f > 0$, we call $\overrightarrow{X^r L} = (X^r, \mathbf{x}_{l+1}^r)$ the *extension* of $X^r \in P(L)$ in case

$$x_{l+1}^r(e) = \begin{cases} 1 \text{ if } e = e_\beta \\ 0 \text{ otherwise.} \end{cases} \tag{5.10}$$

♣

That is, $P^r = (e_4^r, \ldots, e_l^r = e_\alpha)$ and this pedigree can be extended to $(e_4^r, \ldots, e_l^r = e_\alpha, e_\beta)$. And the corresponding characteristic vector, $(X^r, \mathbf{x}_{l+1}^r) \in P_{l+1}$. Observe that $v_f > 0$ implies a generator of e_β is in P^r and e_β does not appear in P^r.

Definition 5.10 Given l, $4 \le l < k \le n$, and $P \in P_l$, $EXT(P, k)$ denotes the set of pedigrees in P_k that are extensions of $P \in P_l$, $4 \le l < k \le n$. That is,

$$EXT(P, k) = \{P' \in P_k | P = P'/l\}. \qquad \clubsuit$$

We are interested in the pedigree packability of any arc flow in F_k. It is easy to see that for any $[k + 1 : e] \in V_{[k-2]}$ with an arc (P, e) connecting a pedigree $P \in R_{k-1}$, we have a flow $f_{P,e} \ge 0$ and this flow is obviously pedigree packable, as the $path(P, e)$ is the pedigree path bringing that flow into e. We wish to check whether this is true for other arcs in F_k as well.

Explicit use of the fact $X \in P_{MI}(n)$ is made in this section to establish the pedigree packability of any node capacity at layer $k - 2$ given that $X/k \in conv(P_k)$.

Theorem 5.2 *Given $X \in P_{MI}(n)$ and (N_{k-1}, R_{k-1}, μ) well-defined, consider any $\lambda \in \Lambda_k(X)$. Let the total flow from pedigrees in R_{k-1} to $[k + 1 : e'] \in V_{[k-2]}$ be denoted by $\delta_{e'}$. For any $[k : e] \in V_{[k-3]}$ and $[k + 1 : e'] \in V_{[k-2]}$, we have a flow f^L in $N_{k-1}(L)$ for a link $L = (e, e')$, such that,*

(1) The value of the flow f^L, given by v^L, is pedigree packable for link L,
(2) For any $e' \in V_{[k-2]}$, we have $\sum_{L=(e,e'),e \in V_{[k-3]}} v^L = x_{k+1}(e') - \delta_{e'}$, and
(3) $\sum_{L=(e,e'),e \in V_{[k-3]}} \sum_{X^r \in \mathcal{P}(L),\, x^r(u)=1} \delta_r \le c(u), u \in \mathcal{V}(N_{k-1})$, where, δ_r is the flow along the $path(X^r)$, $c(u)$ is the capacity for node u. and

$$\mathcal{P}(L) = \{X^r, r \in I(\lambda) \mid x_k^r(e) = 1, e' \in T^r\},$$

where T^r is the tour corresponding to X^r. $\qquad \heartsuit$

In other words, Theorem 5.2 assures the existence of pedigree paths in N_{k-1} bringing in a flow of v^L into the sink, $[k : e]$, and all these paths can be extended to pedigree paths in N_k or in R_k bringing in a total flow of $x_{k+1}(e')$ into $[k + 1 : e'] \in V_{[k-2]}$.

Proof Note that $X/k + 1 \in P_{MI}(k + 1)$. Recall the definition of $U^{(k)}$ in the reformulation of *MI-relaxation* (Problem 4.2). We have,

$$U^{(k)} - A^{(k+1)}\mathbf{x}_{k+1} = U^{(k+1)} \ge 0.$$

Remember that $U^{(k)}$ is the slack variable vector corresponding to X/k. So, $x_{k+1}(e') \le U^{(k)}(e')$. Now $[k + 1 : e'] \in V_{[k-2]}$ means $x_{k+1}(e') > 0$, and so $U^{(k)}(e') > 0$. Since $X/k \in conv(P_k)$, consider any weight vector $\lambda \in \Lambda_k(X)$.

From Lemma 3.1 we have, if X is an integer solution to the MI-relaxation problem for n cities, then U, the corresponding slack variable vector, is the edge-tour incident vector of the n-tour corresponding to X. Applying this for $n = k$ and noticing $X/k \in conv(P_k)$, we find the same λ can be used to write $U^{(k)}$ (the slack variable vector for X/k) as a convex combination of $\dot{T}^r$, $r \in I(\lambda)$, where T^r is the $k - tour$ corresponding to X^r and $\dot{T}$ denotes the edge-tour incident vector of T.

Every X^r corresponding to a pedigree $P^r \in R_{k-1}$ appears in each of these λs with the *fixed weight* $\lambda_r = \mu_{P^r}$.

We have

$$\sum_{X^r \in I(\lambda),\ x^r(u)=1} \lambda_r = c(u),\ u \in \mathcal{V}(N_{k-1}). \tag{5.11}$$

Let

$$J_{(e')} = \{r | X^r \in P_k \text{ and } e' \in T^r\}.$$

Thus,

$$U^{(k)}(e') = \sum_{r \in I(\lambda) \cap J_{e'}} \lambda_r \dot{T}^r. \tag{5.12}$$

Now partition $I(\lambda)$ with respect to $\mathbf{x}_k^r$ as follows: Let I_e denote the subset of $I(\lambda)$ with $x_k^r(e) = 1$.

$$\sum_{r \in I_e} \lambda_r = x_k(e),\ \text{for } e \in E_k.$$

We have,

$$I(\lambda) \cap J_{e'} = \bigcup_{e | I_e \neq \emptyset} I_e \cap J_{e'}. \tag{5.13}$$

As I_e's are disjoint, we a have a partition of $I(\lambda) \cap J_{e'}$. The $path(X^r)$ corresponding to $r \in I_e \cap J_{e'}$, is available in $N_{k-1}(L)$, since any $X^r \in P_k$ active for X/k is such that the $path(X^r)$ is available in N_{k-1} or it corresponds to a rigid pedigree (see Lemma 5.5).

Now let

$$\mathcal{P}(L) = \{X^r | r \in I_e \cap J_{e'}\}.$$

Or equivalently,

$$\mathcal{P}(L) = \{X^r, r \in I(\lambda) \mid x_k^r(e) = 1, e' \in T^r\}.$$

Any $path(X^r)$ for an $X^r \in \mathcal{P}(L)$ can be extended to a path ending in $[k + 1 : e']$, using the arc $([k : e], [k + 1 : e'])$. Thus we have a subset of pedigrees in P_{k+1}, corresponding to these extended paths. We see from Eqs. 5.12 and 5.13 that we can do this for each e, and a maximum of $U^{(k)}(e')$ can flow into $[k + 1 : e']$.

Of which, along the rigid pedigrees in R_{k-1}, we have a total flow of $\delta_{e'}$ into $[k + 1 : e']$. Since $x_{k+1}(e') \le U^{(k)}(e')$, we can choose nonnegative $\delta_r \le \lambda_r$, so that we have exactly a flow of $x_{k+1}(e') - \delta_{e'}$ into $[k+1 : e']$ along the paths corresponding to non-rigid pedigrees $X^r \in \cup_e \mathcal{P}(L)$. Now we have part 3 of the theorem, from

(a) $\cup_e \mathcal{P}(L)$ is a subset of $\{X^r | r \in I(\lambda)\}$,
(b) $\delta_r \leq \lambda_r$ and
(c) the expression for $c(u)$, given by Eq. 5.11.

Letting $v_L = \sum_{X^r \in \mathcal{P}(L)} \delta_r$ we have the parts 1 and 2 of the result. Hence the theorem.
□

Remark Even though we can apply this theorem for any e' with $x_{k+1}(e') > 0$, the simultaneous application of this theorem for more than one e', in general, may not be correct. This is so because, for some paths the total flow with respect to the different e' may violate the node capacity for a node at some layer, l. Example 5.4 illustrates this point.

Corollary 5.1 *Given* $X \in P_{MI}(n)$ *and* $X/k \in conv(P_k)$, *if* $x_{k+1}(e') = 1$ *for some* e', *then* $X/k + 1 \in conv(P_{k+1})$. ♡

Proof $X/k + 1$ as given, means that e' is available for insertion of $k + 1$ with certainty. In other words, every pedigree active for X/k is such that the corresponding $k - tour$ contains e'. Essentially, the proof lies in seeing the fact that given a $\lambda \in \Lambda_k(X)$, we can extend every $X^r, r \in I(\lambda)$ to $(X^r, ind(e'))$ with the same weight λ_r, where $ind(e')$ is the indicator of e'.

Now refer to the proof of Theorem 5.2. Since $x_{k+1}(e') = 1$, we have $x_{k+1}(e') = U^{(k)}(e')$. So $\delta_r = \lambda_r, r \in I(\lambda) \cap J_{e'}$. It follows from an application of Theorem 5.2 and the above observation that the part 3 of the theorem yields strict equality for each node in each layer of N_{k-1}. This with part 2 of the theorem implies the required result.

Corollary 5.2 *Given* $X \in P_{MI}(n)$ *and* $X/k \in conv(P_k)$, *if* x_{k+1} *is such that* $x_{k+1}(e) = 0, e \in E_{k-1}$, *that is,*

$$\sum_{i=1}^{k-1} x_{k+1}((i, k)) = 1, \tag{5.14}$$

then F_k *feasible implies* $X/k + 1 \in conv(P_{k+1})$. ♡

Proof Since F_k is feasible consider any flow $f = (f_{u,v})$, $(u, v) \in \mathcal{A}$. Consider any origin u such that either $u = [k : e = (i, j)]$ or $u = P$ and P ends in e for some $1 \leq i < j < k$. We may have flows from u to $v = [k + 1 : (i, k)]$ or $v' = [k + 1 : (j, k)]$ only, as other arcs are forbidden or don't exist under the given condition on $\mathbf{x}_{k+1}$. Since $X/k \in conv(P_k)$, consider any $\lambda \in \Lambda_k(X)$. Pedigrees in $I(\lambda)$ can be partitioned depending on their last component e. The availability at u is either $\bar{x}_k(e')$ or $\bar{\mu}_P$ and is equal to $f_{u,v} + f_{u,v'}$ as F_k is feasible. Thus these pedigrees can be extended to pedigrees in P_{k+1} with weights $f_{u,v}$, $f_{u,v'}$ respectively. We can do this for all u. Hence we have expressed $X/k + 1$ as a convex combination of pedigrees in P_{k+1}, as required.

Remark

1. Corollary 5.1 is useful in proving Theorem 5.4, which is an important result in this book.
2. In Corollary 5.2, satisfying the condition given by Eq. 5.14 does not automatically guarantee the feasibility of F_k, as can be seen from Example 5.5.
3. The Corollaries 5.1 and 5.2 are useful in concluding $X/k + 1 \in conv(P_{k+1})$ with easy to check conditions. In general we need to solve a multicommodity flow problem to declare (N_k, R_k, μ) is well-defined, as shown in Sect. 5.5.

Example 5.4 Consider X as given below:

$$\mathbf{x}_4 = (0, 3/4, 1/4);$$
$$\mathbf{x}_5 = (1/2, 0, 0, 0, 0, 1/2);$$
$$\mathbf{x}_6 = (1/8, 1/8, 3/8, 0, 1/8, 2/8, 0, 0, 0, 0).$$

It can be verified that $X \in P_{MI}(6)$, and $X/5 \in conv(P_5)$ as F_4 is feasible. See Fig. 5.6a. As can be seen from Fig. 5.6b flow along each link L is pedigree packable. Also, F_5 is feasible. Arcs $[5 : 3, 4] \rightarrow [6 : 1, 2]$ and $[5 : 1, 2] \rightarrow [6 : 3, 4]$ are rigid with frozen flows, $1/8$ and $2/8$ respectively. See Fig. 5.6c. However, it is not obvious whether $X \in conv(P_6)$, or not. Figure 5.7 gives the pedigree paths with flows, and it can be checked that these flows saturate all node capacities. Hence, $X \in conv(P_6)$.

But this may not be the case in general. For instance, consider X^* given by

$$\mathbf{x}^*_4 = \mathbf{x}_4;$$
$$\mathbf{x}^*_5 = \mathbf{x}_5;$$
$$\mathbf{x}^*_6 = (0, 1/4, 1/2, 0, 1/4, 0, 0, 0, 0, 0).$$

It can be checked, F_5 is feasible but $X/6 \notin conv(P_6)$.
Instead, consider X' as given below:

$$\mathbf{x}'_4 = (0, 3/4, 1/4);$$
$$\mathbf{x}'_5 = (1/2, 0, 0, 1/2, 0, 0);$$
$$\mathbf{x}'_6 = (0, 0, 0, 0, 0, 1, 0, 0, 0, 0).$$

[a] F_4 is feasible and no arcs are Rigid.

Restricted Networks for Links			
$N_4(L)$	L	C(L)	Unique Path?
[4:1,3] -> [5:3,4] [4:2,3] -> [5:3,4]	[5:3,4] -> [6:1,2]	1/2	No
[4:2,3] -> [5:1,2]	[5:1,2] -> [6:1,3]	1/4	Yes
[4:2,3] -> [5:3,4]	[5:3,4] -> [6:1,3]	1/4	Yes
[4:1,3] -> [5:1,2]	[5:1,2] -> [6:2,3]	1/2	Yes
[4:1,3] -> [5:3,4]	[5:3,4] -> [6:2,3]	1/2	Yes
[4:2,3] -> [5:1,2]	[5:1,2] -> [6:2,4]	1/4	Yes
[4:2,3] -> [5:3,4]	[5:3,4] -> [6:2,4]	1/4	Yes
[4:1,3] -> [5:1,2] [4:2,3] -> [5:1,2]	[5:1,2] -> [6:3,4]	1/2	No

[b] Restricted Networks for Links

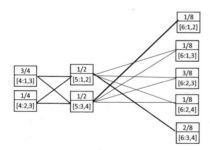

[c] F_5 is feasible. The rigid arcs are shown with thick lines.

Fig. 5.6 Networks for Example 5.4, **a** N_4, **b** $N_4(L)'s$, and **c** N_5

We can check that the paths

$$[4:2,3] \rightarrow [5:1,2] \rightarrow [6:3,4],$$

$$[4:1,3] \rightarrow [5:1,2] \rightarrow [6:3,4],$$

and

$$[4:1,3] \rightarrow [5:1,4] \rightarrow [6:3,4]$$

Fig. 5.7 Pedigree paths
showing $X \in conv(P_6)$ for
Example 5.4

Pedigree Path	flow
[4:1,3]→[5:3,4]→[6:1,2]	1/8
[4:2,3]→[5:1,2]→[6:1,3]	1/8
[4:1,3]→[5:3,4]→[6:2,3]	2/8
[4:1,3]→[5:1,2]→[6:2,3]	1/8
[4:2,3]→[5:3,4]→[6:2,4]	1/8
[4:1,3]→[5:1,2]→[6:3,4]	2/8

bring the flows of $1/4$, $1/4$ and $1/2$, respectively to $[6:3,4]$, adding to $1 = x_6(3,4)$.
And so $X' \in conv(P_6)$, as assured by Theorem 5.2.

Example 5.5 Consider X as given below:

$$\mathbf{x}_4 = (0, 1/2, 1/2);$$
$$\mathbf{x}_5 = (0, 0, 0, 0, 0, 1);$$
$$\mathbf{x}_6 = (1/2, 0, 0, 0, 0, 0, 0, 0, 1/2, 0)$$
$$\mathbf{x}_7 = (0, 0, 0, 0, 0, 0, 0, 0, 0, 0, 1/4, 0, 1/2, 0, 1/4).$$

We can check that X is in $P_{MI}(7)$, and the paths

$$[4:2,3] \rightarrow [5:3,4] \rightarrow [6:1,2],$$

$$[4:1,3] \rightarrow [5:3,4] \rightarrow [6:3,5],$$

bring a flow of $1/2$ to $[6:1,2]$, and $[6:3,5]$ respectively. And so $X/6 \in conv(P_6)$.
Though $\mathbf{x}_7$ satisfies the Eq. 5.14, F_6 is not feasible, as the total demand at
$[7:3,6]$ & $[7:5,6]$ is $3/4$ and only $[6:3,5]$ with availability $1/2$ is connected
to these nodes.

5.5 A Multicommodity Flow Problem to Check
     Membership

Our next task is to check that (N_k, R_k, μ) is well-defined. To declare the network (N_k, R_k, μ) to be well-defined, we need to have evidence that $X/k+1 \in$
$conv(P_{k+1})$. This was easy for $k = 4$. As seen in the Example 5.3, even though there
are pedigree paths bringing the flow along each arc in F_k, there could be conflicts arising out of the simultaneous capacity restrictions on these flows in the network N_{k-1}.
This calls for ensuring that these restrictions are not violated. The multicommodity
flow problem defined in this section does precisely this check.

5.5.1 Defining the Multicommodity Flow Problem

Definition 5.11 (*Commodities*) Consider the network (N_k, R_k, μ). For every arc $a \in \mathcal{A}(N_k) \setminus \mathcal{A}(N_4)$ designate a unique *commodity* s. Let S denote the set of all commodities. We write $a \leftrightarrow s$ and read a designates s. Let S_l denote the set of all commodities designated by arcs in F_l. ♣

Definition 5.12 Given a pedigree $X^r \in P_{k+1}$, we say that it agrees with an arc $a = (u, v) \in F_l, 4 \leq l \leq k$ in case either (1) $u = [l : e_l^r]$, $v = [l + 1 : e_{l+1}^r]$ or (2) $u = P^r/l \in R_{l-1}$ with $v = [l + 1 : e_{l+1}^r]$. We denote this by, $X^r \parallel a$ and read X^r agrees with a. For any $\lambda \in \Lambda_{k+1}(X)$ and $r \in I(\lambda)$, let $I_s(\lambda)$ denote the subset of $I(\lambda)$ such that the corresponding pedigrees agree with arc a which designates s. ♣

Figure 5.8 explains the concepts (1) a pedigree 'agreeing' with an arc and (2) an arc 'designating' a commodity.

Next, we define the multicommodity flow problem.

Let c_a be the capacity of any arc a in the network. Let $f_a, f_a^s \geq 0$ be the flow through arc a and the flow of commodity s through arc a respectively. For any s, we allow this flow to be positive only for the arcs in the restricted network corresponding to s, denoted by $N_{l-1}(s)$, for $s \in S_l$. For each $l, 5 \leq l \leq k$, we have the following capacity restrictions on the flow through any arc:

$$c_a \geq f_a \geq 0, \ a \in \mathcal{A}(N_k) \tag{5.15}$$

$$f_a^s \leq \begin{cases} c_a & \text{if } a \in \mathcal{A}(N_{l-1}(s)) \\ 0 & \text{otherwise.} \end{cases} \tag{5.16}$$

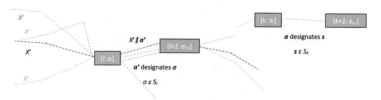

Legend: [1] Different pedigree paths are shown in different colours. The pedigrees X^r are shown in matching colour.
[2] Some of these pedigrees agree with an arc a' and they contribute to $f_{a'}$, like the black, orange, and blue pedigrees.
[3] Pedigrees that agree with an arc a designating commodity s contribute to the corresponding commodity flow, like the orange pedigree contributing to commodity flow for s. [4] Orange pedigree agrees with a' so contributes to flow of s along arc a'.

Fig. 5.8 Pedigrees $X^r \parallel a'$, arc a designating s

For each commodity, s at each intermediate node v we conserve the flow:

$$\sum_{u \,\ni\, a=(u,v)} f_a^s \;=\; \sum_{w \,\ni\, a=(v,w)} f_a^s, \; v \in \mathcal{V}(N_{l-1}) \setminus \{V_{[1]} \cup V_{[l-2]}\}, s \in \mathcal{S}_l. \quad (5.17)$$

We require the total flow of all commodities in $\mathcal{S}_l$ through an arc to be the flow f_a:

$$\sum_{s \in \mathcal{S}_l} f_a^s = f_a, \; a \text{ an arc in } \mathcal{A}(N_{l-1}). \quad (5.18)$$

We define the total flow of commodity s through the arc a designating s to be v_s and is equal to the flow along the arc:

$$\sum_{a' \in \delta^-(a)} f_{a'}^s = f_a \triangleq v_s, a \leftrightarrow s, s \in \mathcal{S}_l, \quad (5.19)$$

where $\delta^-(a)$ is the set of arcs entering a. In addition, at each node $v \in \mathcal{V}(N_l)$ we have the node capacity restriction on the total flow through the node as well:

$$\sum_{s \in \mathcal{S}_l} \sum_{u \ni a=(u,v)} f_a^s \leq \bar{x}(v), \; v \in \mathcal{V}(N_l). \quad (5.20)$$

We also have the availability restrictions at the source nodes in $V_{[1]}$ and in R_q:

$$\sum_{w \ni a=(u,w)} f_a \leq \bar{x}(u), u \in V_{[1]}. \quad (5.21)$$

$$\sum_{w \ni a=(u,w)} f_a \leq \bar{\mu}_u, u \in R_q, q \in 4 \leq q \leq l-1. \quad (5.22)$$

We also have the node capacity restrictions at the nodes in $V_{[l-2]}$:

$$\sum_{u \ni a=(u,v)} f_a \leq \bar{x}(v), v \in V_{[l-2]}. \quad (5.23)$$

The objective is to maximise the total flow along the arcs in F_k:

$$z = \sum_{a \in F_k} f_a = \sum_{s \in \mathcal{S}_k} v^s. \quad (5.24)$$

The multicommodity flow problem now can be stated as:

Problem 5.1 Problem: $MCF(k)$

$$maximise \ z = \sum_{s \in S_k} v^s \tag{5.25}$$

$$subject \ to \tag{5.26}$$
$$for \ all \ \ l, 5 \leq l \leq k$$
$$constraints \ \ (5.15) \ through \ (5.23).$$

Next, we observe some easy to prove facts about the multicommodity network and Problem 5.1.

Let z^* denote the objective function value for an optimal solution to Problem 5.1. Let $f = (f^1, \ldots, f^{|S|})$ denote any feasible multicommodity flow for Problem 5.1, where f^s gives the flow vector for commodity $s \in S_l$, for a fixed ordering of the arcs in the network, $N_{l-1}, 5 \leq l \leq k$.

Remark

1. z^* can be at most $z_{max} = 1 - \sum_{P \in R_k} \mu_P$.
2. If $z^* = z_{max}$, then for any optimal solution to Problem 5.1, the bundle capacity $\bar{x}(v)$ at each node $v \in V_{[l-3]}, 4 \leq l \leq k+1$ is saturated.
3. Every feasible path bringing a positive flow for any commodity $s \in S_k$ passes through each layer of the network, satisfying the commodity flow restrictions for s.
4. If $z^* = z_{max}$ then any optimal solution to Problem 5.1, is such that the solution restricted to the portion of the network corresponding to arcs in F_l, constitutes a feasible solution to F_l. This follows from item [2] of this remark. Now, any such flow saturates the node capacities at layers $l - 3$ and $l - 2$ and $f_a^s, a \in F_l$ can be positive only for the arc a designating s. So letting $\sum_{s \in S_l} f_a^s = g_a$, we can check that g is feasible for F_l as all the restrictions of problem F_l are also present in Problem 5.1.

 Note *Problem 5.1 is defined to prove Theorems 5.3 and 5.4, that give a necessary and sufficient condition for the membership of X in the pedigree polytope.*

5.5.2 Proving the Necessity of the Condition

Next we prove the necessity of the multicommodity flow problem having a feasible solution with an optimum of z_{max}, for membership of $X/k + 1$ in $conv(P_{k+1})$.

Theorem 5.3 *Given $X/k + 1 \in conv(P_{k+1})$ then there exists f, f^s, feasible for the multicommodity flow problem (Problem 5.1), with $z^* = \sum_{s \in S_k} v^s = z_{max}$.* ♡

Proof Since $X/k + 1$ is in $conv(P_{k+1})$ we have from the proof of Theorem 5.1, for any $\lambda \in \Lambda_{k+1}(X)$, a feasible solution to F_l is given by the instant flow for $INST(\lambda, l)$. Define f_a as follows:

$$f_a = \begin{cases} \sum_{r \in I(\lambda)|X^r \parallel a} \lambda_r & \text{if } a \in N_k \\ 0 & \text{otherwise.} \end{cases} \tag{5.27}$$

For $5 \le l \le k$ and $s \in S_l$ define f_a^s as follows:

$$f_a^s = \begin{cases} \sum_{r \in I_s(\lambda)|X^r \parallel a} \lambda_r & \text{if } a \in N_{l-1}(s) \\ 0 & \text{otherwise.} \end{cases} \tag{5.28}$$

We shall show that f, f^s as defined are feasible for Problem 5.1 and the objective value is z_{max}. Non negativity restrictions on f_a and f_a^s are met as λ_r are positive for all $r \in I(\lambda)$. Since the instant flow for $INST(\lambda, l)$ is feasible for F_l, all the capacity restrictions on arcs and nodes are met.

For each commodity $s \in S_l$, for $r \in I_s(\lambda)$, note that X^r/l represents a path in $N_{l-1}(s)$ or the corresponding path is in R_{l-1}. For any of these paths, commodity flow is conserved at each node along the path. And a node not in any of these paths, does not have a positive flow of this commodity through that node.

We have,

$$\sum_{u \, \ni \, a=(u,v)} f_a^s = \sum_{u \, \ni \, a=(u,v)} \sum_{r \in I_s(\lambda)|X^r \parallel a} \lambda_r \tag{5.29}$$

$$= \sum_{r \in I_s(\lambda)|x_l^r(e)=1} \lambda_r, \tag{5.30}$$

for $v = [l : e] \in \mathcal{V}(N_l), s \in S_l$.

Similarly

$$\sum_{w \, \ni \, a=(v,w)} f_a^s = \sum_{w \, \ni \, a=(v,w)} \sum_{r \in I_s(\lambda)|X^r \parallel a} \lambda_r \tag{5.31}$$

$$= \sum_{r \in I_s(\lambda)|x_l^r(e)=1} \lambda_r, \tag{5.32}$$

for $v = [l : e] \in V(N_l), s \in S_l$.

Hence, the commodity flow conservation restrictions are all met. Moreover, for $l = k$, the flow v^s along the arc a in F_k defining commodity s in S_k are nonnegative and they add up to $\sum_{v \in V_{[k-2]}} \bar{x}(v)$. And every $\lambda \in \Lambda_{k+1}(X)$ is such that, there exists $r \in I(\lambda)$ corresponding to $P \in R_k$ with $\lambda_r = \mu_P$. Hence, $1 - \sum_{P \in R_k} \mu_P$ is maximum possible flow in Problem 5.1 and equals $\sum_{v \in V_{[k-2]}} \bar{x}(v)$. Hence the total flow in the network is $\sum_{s \in \mathcal{S}_k} v^s = \sum_{s \in \mathcal{S}_k} f_a^s = \sum_{s \in \mathcal{S}_k} \sum_{r \in I_s(\lambda)} \lambda_r = z_{max}$.

Thus we have verified that f, f^s are feasible and the objective function value is z_{max}.

An implication of Theorem 5.3 is what we can conclude when the Problem 5.1 has a maximal flow less than z_{max}, which is stated as Corollary 5.3.

Corollary 5.3 *If Problem 5.1 has a maximal flow $z^* < z_{max}$ then $X/k+1 \notin conv(P_{k+1})$.* ♡

In addition, it is essential to know what can be said about the converse of Theorem 5.3. Towards this end, I shall prove some results in the following subsection.

5.5.3 Proving the Sufficiency of the Condition

First, I prove two lemmas that are used in showing the converse of Theorem 5.3. Assume that we have an optimal solution (f, f^s) for the muticommodity flow problem with $z* = z_{max}$. Recall that any $a \in \mathcal{A}(F_l)$ is such that, either $a = [u, v]$, $u \in V_{[l-3]}$, with $v \in V_{[l-2]}$ or $a = [P, v]$, with $[P, v]$ not rigid, $P \in R_{l-1}$, $v \in V_{[l-2]}$. (If $[P, v]$ were rigid then the extended pedigree would be included in R_l and the arc a will not be included in $\mathcal{A}(F_l)$). Considering the different commodities $s \in \mathcal{S}_k$, when we have a solution to the multicommodity flow problem, essentially we are apportioning $x(u = [l : e])$ to the different commodities that flow through the node u. However, u may be a node in layer $l - 3$ of N_k, if it has not been deleted because the updated capacity became zero, and it may occur in some of the rigid paths corresponding to pedigrees in $R_{l-1}, \ldots, R_{k-1}$.

Definition 5.13 For any $s \in \mathcal{S}_k$, let $Y^s \in R^{\tau_k}$, denote the vector $(y_4^s((1, 2)), \ldots, y_k^s((k - 2, k - 1)))$, with $y_l^s(e)$, $4 \leq l \leq k$, $e \in E_{l-1}$ given by

$$y^s(u) \equiv y_l^s(e) = \begin{cases} \sum_{a=[u,v] \in F_l} f_a^s + \sum_{q=l}^{k} \sum_{a \in \mathcal{A}(u,q)} f_a^s & \text{if } u = [l, e] \in V_{[l-3]}, \\ 0 & \text{otherwise.} \end{cases}$$

(5.33)

where $\mathcal{A}(u, q) = \{a \mid a = [P, v] \in F_q, P \in R_{q-1} \& u \text{ occurs in } \in path(P)\}$. ♣

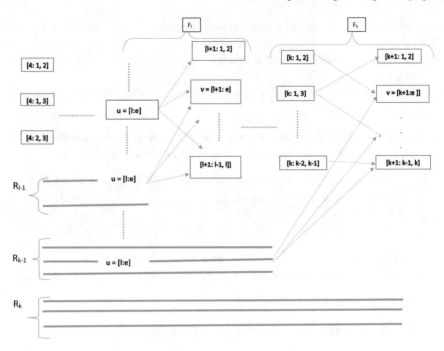

Fig. 5.9 Figure showing the set of Rigid pedigrees that contain u

$A(u, q)$ is the set of all arcs $[P, v] \in F_q$ such that P is a rigid pedigree in R_{q-1}, and the $path(P)$ has u in it. Thus, $\sum_{q=l}^{k} \sum_{a \in A(u,q)} f_a^s$ gives the total flow of commodity s along the rigid paths that contain u. So, $y^s(u)$ gives the capacity used up for the flow of commodity s through node u, as per solution f^s (see Fig. 5.9).

Note that, $\sum_{u \in V_{[l-3]}} y^s(u) = \sum_{e \in E_{l-1}} y_l^s(e) = v^s, 4 \le l \le k$. Moreover, $\sum_{s \in S_k} y^s(u) = \bar{x}(u) = \bar{x}_l(e), 4 \le l \le k, e \in E_{l-1}$. Therefore $\sum_{s \in S_k} Y^s = \bar{X}/k$. [Recall that, $\bar{X}/k$ gives the vector of residual capacities at the nodes, after adjusting for the fixed flow μ_P for rigid pedigree $P \in R_k$.]

Lemma 5.6 *Given that we have an optimal solution for Problem 5.1 with $z^* = z_{max}$, consider the optimal commodity flow $f^s, s \in S_k$. Let Y^s be as defined above in Eq. 5.33, then, $1/v^s(Y^s) \in P_{MI}(k)$, for any $s \in S_k$.* ♡

Proof Let $Y = 1/v^s(Y^s), s \in S_k$. Feasibility of Y for $P_{MI}(k)$ can be verified by checking that we have $U \ge 0$, satisfying the conditions of Lemma 4.5, namely, Eqs. 4.10 through 4.14 are satisfied with $n = k$.

Non negativity of Y is met by definition. And the constraints $E_{[k]}Y = \mathbf{1}_{k-3}$ in Eq. 4.10 are easily met by Y, as $\sum_{e \in E_{l-1}} y_l^s(e) = v^s$, $4 \le l \le k$. Recall that $U^{(3)} = \begin{pmatrix} \mathbf{1}_3 \\ \mathbf{0} \end{pmatrix}$, and $U^{(l)}$ satisfies:

$$U^{(l)} - A^{(l+1)} y_{l+1} = U^{(l+1)}, \ 3 \le l \le k \tag{5.34}$$

(See Chap. 4 for definitions.) We shall check that $U^{(l+1)} \ge 0$ for $3 \le l \le k$.

For $l = 3$ as $y_4^s(e) \ge 0$, $e \in E_3$, $\sum_{e \in E_3} y_4^s(e) = v^s$. So $y_4(e) = 1/v^s(y_4^s(e)) \le 1$, $e \in E_3$ and so $U^{(4)} \ge 0$.

Suppose the result is not true for all $l > 3$, then there exists a $q \ge 4$ for which, $U_{i^o j^o}^{(q+1)} < 0$ for some $e = (i^o, j^o)$.

Since $Y^s \le X/k$, the support of X/k also supports Y (however, some $y_l^s(e)$ could be 0). Now consider the construction of the network N_q. In the network (N_{q-1}, R_{q-1}, μ), the maximum possible flow of commodity s into $[q : e]$ can not be more than the flow through the generator nodes of $e = (i^o, j^o)$. That is, the maximum flow is less than or equal to $\sum_{e' \in G(e)} y_{j^o}^s(e')$. Moreover, out of this, nodes $[l : (i^o, j^o)]$, $j^o + 1 \le l \le q - 1$ receive $\sum_{l=j^o+1}^{q-1} y_l^s(e)$. But no part of this flow of commodity s can flow into $[q : e]$ with $e = (i^o, j^o)$. Thus the maximum possible flow is less than or equal to $\sum_{e' \in G(e)} y_{j^o}^s(e') - \sum_{l=j^o+1}^{q-1} y_l^s(e)$, which by definition is $U_{i^o j^o}^{(q)}$ and is nonnegative. On the other hand, $y_q^s(e)$ by definition is the total flow of commodity s passing through node $[q : e]$ and so it cannot be larger than $U_{i^o j^o}^{(q)}$ as the flow conservation is met for commodity s at each intermediate node of $N_{k-1}(s)$. So, we have $U_{i^o j^o}^{(q+1)} \ge 0$, leading to a contradiction. □

Next we shall show that for $k = 5$ we have Lemma 5.7, which will form the basis for a proof by induction to show that $1/v^s(Y^s/k) \in conv(P_k)$, $s \in S_k$.

Lemma 5.7 Let $k = 5$, and σ be a commodity in S_5, designated by $a = ([5 : e_\sigma'], [6 : e_\sigma])$. If the multicommodity flow problem (Problem 5.1), is feasible with $z^* = \sum_{\sigma \in S_5} v^\sigma = z_{max}$, then

1. $1/v^\sigma(Y^\sigma) \in conv(P_5)$, $\sigma \in S_5$, and so
2. $(1/v^\sigma(Y^\sigma), ind(e_\sigma)) \in conv(P_6)$, $\sigma \in S_5$. Hence, $X/6 \in conv(P_6)$.   ♡

Proof If the conditions of the lemma are satisfied, then $z_{max} = 1 - \sum_{P \in R_4} \mu_P$. If $N_4 = \emptyset$, there is nothing to prove as all commodity flows are using rigid pedigree paths. If $N_4 \ne \emptyset$, for any $\sigma \in S_5$, all the commodity flow paths in $N_4(\sigma)$ are pedigree paths as they correspond to arcs in N_4. Hence, we have statement [1] of the lemma. Since any such pedigree path ends in the $tail(a)$, where $a \leftrightarrow \sigma$, it can be extended to a pedigree path corresponding to a pedigree in P_6 using $head(a)$. And we can do this for each $\sigma \in S_5$. As $\sum_{\sigma \in S_5} v^\sigma(1/v^\sigma(Y^\sigma), ind(e_\sigma)) = X/6$, we have $X/6 \in P_6$ proving statement [2].

Remark Observe that for $k > 5$, $1/v^s(Y^s/5) \in conv(P_5)$, $s \in \mathcal{S}_k$. (similar to the proof of statement [1]). Now $\sum_{s \in \mathcal{S}_k} f_a^s = v^\sigma$, $a \leftrightarrow \sigma$. Therefore, for each $s \in \mathcal{S}_k$ and $a \in F_5$, f_a^s is pedigree packable. So $1/v^s(Y^s/6) \in conv(P_6)$.

Lemma 5.8 *Given that we have an optimal solution for MCF(k) (Problem 5.1)) with $z^* = z_{max}$, consider the optimal commodity flow f^s, $s \in \mathcal{S}_k$. Let Y^s be as defined in Eq. 5.33, then, $1/v^s(Y^s/k) \in conv(P_k)$, $s \in \mathcal{S}_k$.*   ♡

Proof *(By induction)* Basis: $[l = 5]$ We have $1/v^s(Y^s/5) \in conv(P_5)$, $s \in \mathcal{S}_k$, from the remark following Lemma 5.7.

Induction hypothesis: Assume that the result is true for all $l \leq k - 1$, that is, $1/v^s(Y^s/l) \in conv(P_l)$, for $s \in \mathcal{S}_k$.

We shall show that the result is true for $l = k$. Now an application of Theorem 5.2, with the facts [a] $1/v^s(Y^s/k - 1) \in conv(P_{k-1})$ (induction hypothesis for $l = k - 1$) and [b] $1/v^s(Y^s/k) \in P_{MI}(k)$ from Lemma 5.6, we have pedigree packability of any arc a' in F_{k-1} that has a flow $f_{a'}^s$. However, we need to show that all such arcs in F_{k-1} carrying the flow for commodity s are simultaneously pedigree packable. But commodity s is carried only by arcs $[w, u]$ ending in $u = [k : e']$, for some e' and arc $[u, v]$ in F_k designates s. This implies $y_k^s(e') = \sum_{a'=[w,u]} f_{a'}^s = v^s$. Therefore, $1/v^s(Y^s/k)$ has the last component as a unit vector corresponding to e'. Thus, we have conditions of Corollary 5.1 satisfied. Hence, $1/v^s(Y^s/k) \in conv(P_k)$ as required.

Theorem 5.4 *If the multicommodity flow problem (Problem 5.1), is feasible with $z^* = \sum_{s \in \mathcal{S}_k} v^s = z_{max}$, then $X/k + 1 \in conv(P_{k+1})$.*   ♡

Proof Assume that the multicommodity flow problem $MCF(k)$, is feasible with $z^* = \sum_{s \in \mathcal{S}_k} v^s = z_{max}$. Let $a = ([k : e_s'], [k + 1 : e_s]) \leftrightarrow s$. From Lemma 5.8 we have $1/v^s(Y^s/k) \in conv(P_k)$, $s \in \mathcal{S}_k$. Consider any such s, there exists $\gamma_P > 0$, and $\sum_{P \in \mathcal{P}^s} \gamma_P = 1$, such that $1/v^s(Y^s/k) = \sum_{P \in \mathcal{P}^s} \gamma_P X_P$, where $\mathcal{P}^s$ is a subset of pedigrees in P_k, and X_P is the characteristic vector of P. Each such P ends in e_s'. Therefore, P can be extended to a pedigree in P_{k+1} using e_s. Notice that the weight of any such P is $v^s \gamma_P$. And these weights add up to v^s. We can do this for each $s \in \mathcal{S}_k$. Thus we have shown that $\sum_{s \in \mathcal{S}_k} 1/v^s(Y^s, ind(e_s)) \in conv(P_{k+1})$. But $\sum_{s \in \mathcal{S}_k} v^s(1/v^s(Y^s, ind(e_s))) = \bar{X}/k + 1$. Now all $P \in R_k$ are pedigrees in P_{k+1}. As $\sum_{s \in \mathcal{S}_k} v^s = z_{max} = 1 - \sum_{P \in R_k} \mu_P$, and $\bar{X}/k + 1 + \sum_{P \in R_k} \mu_P X_P = X/k + 1$, we have $X/k + 1 \in conv(P_{k+1})$. Hence the theorem.

This is an important converse result, which shows that having a feasible solution with an objective function value equal to z_{max} for $MCF(k)$ (Problem 5.1) is sufficient for membership in the relevent pedigree polytope.

Thus, I can state the main result of this chapter giving a necessary and sufficient condition for membership in the pedigree polytope as Theorem 5.5.

Theorem 5.5 *Given n, $X \in P_{MI}(n)$, and $X/n - 1 \in conv(P_{n-1})$. Then $X \in conv(P_n)$ if and only if there exists a solution for the multicommodity flow problem MCF(n) (Problem 5.1 with $k = n$), with $z^* = \sum_{s \in \mathcal{S}_n} v^s = z_{max}$.* ♡

Remark This remark is on the earlier necessary and sufficient condition characterising $conv(P_n)$ given by Theorem 4.4, as well as that given now by Theorem 5.5.

1 Recall that we do have a necessary and sufficient condition for membership in the pedigree polytope given in Sect. 4.4, that is, Theorem 4.4. Unfortunately, there was no discussion there or elsewhere on the complexity of finding such a λ as stated in Theorem 4.4.
2 Therefore, the importance of the main result of the present chapter (Theorem 5.5) enhances if we can assess the complexity of testing the newly found necessary and sufficient condition stated in Theorem 5.5.
3 We can however make use of Theorem 4.4 as follows: If for some $l > 5$ if $(N_{l-2} = \emptyset, R_{l-2}, \mu)$ is well-defined and F_{l-1} is feasible, we can conclude $X/l \in conv(P_l)$. This is so because, $FAT_{l-1}(\lambda)$ problem of the Theorem 4.4 coincides with F_{l-1}. So, we need not go for solving a multicommodity flow problem. We simply construct (N_{l-1}, R_{l-1}, μ) and declare it well-defined and proceed further.

The next chapter is devoted to estimating the computational complexity of checking this necessary and sufficient condition given by Theorem 5.5.

Chapter 6
Computational Complexity of Membership Checking

Implementing the Suggested Framework

6.1 Computational Complexity of Checking the Necessary and Sufficient Condition

In this section, we explore how easy it is to check the condition given by the main theorem in Chap. 5 for $X/k + 1$ to be in $conv(P_{k+1})$. Properties of the pedigree polytope known are useful in carrying out this estimation. Especially, we require [1] the dimension of the pedigree polytope and [2] the geometry of the adjacency structure of the polytope; and we recall results from Chap. 4 and other related works [11, 14, 16].

6.1.1 On the Mutual Adjacency of Pedigrees in the Set of Rigid Pedigrees

In this subsection, given $k \geq 5$, and (N_{k-1}, R_{k-1}, μ), is well-defined, we shall show an important theorem on the mutual adjacency in $conv(P_k)$ of pedigrees in R_{k-1}. In addition, we discuss how the size of the set of rigid pedigrees, $|R_k|$ grows as a function of k. In general, if either $R_{k-1} = \emptyset$ or a singleton, we have nothing to prove. Assume without loss of generality $|R_{k-1}| > 1$.

Lemma 6.1 *First we notice that for $k = 5$ when (N_{k-1}, R_{k-1}, μ) is well-defined, R_4 can not have $P^{[i]} = (e_4^i, e_5^i), i = 1, 2$, pedigrees from $\mathcal{P}_5$ that are nonadjacent. Or equivalently, pedigrees in R_4 are mutually adjacent.* ♡

T. S. Arthanari, *Pedigree Polytopes*,
https://doi.org/10.1007/978-981-19-9952-9_6

Proof Suppose $P^{[i]} = (e_4^i, e_5^i), i = 1, 2$, are nonadjacent in the pedigree polytope[1] for $k = 5$. Since, $conv(P_k), k > 3$ are combinatorial polytopes (see Chap. 4 for a proof) [11], from the definition of combinatorial polytopes (see Sect. 2.6 for definition) we have two other pedigrees $P^{[3]}, P^{[4]} \in P_5$ such that, $X^* = 1/2(X^{[1]} + X^{[2]}) = 1/2(X^{[3]} + X^{[4]})$, where $X^{[i]}$, is the characteristic vector of the corresponding pedigree, $P^{[i]}, i = 1, \ldots, 4$. Since the support of X^*, supports $X^{[3]}$, as well as $X^{[4]}$, one of these (without loss of generality say $X^{[3]}$) will have the first component of $X^{[1]}$, and the second component, different from that of $X^{[1]}$ and so necessarily, it is the second component of $X^{[2]}$. Similarly $X^{[4]}$ has the first component of $X^{[2]}$, and the second component of $X^{[1]}$. Thus, we can have a positive flow, $\epsilon < min(\mu(P^{[1]}), \mu(P^{[2]}))$, along the arcs for $P^{[3]}$ and $P^{[4]}$ and decrease the same along the arcs for $P^{[1]}$ and $P^{[2]}$ in F_4, this contradicts the rigidity of the arcs for $P^{[1]}$ and $P^{[2]}$. Hence any two $P^{[i]} \in R_{k-1}, i = 1, 2$, are adjacent in the pedigree polytope for $k = 5$, namely $conv(P_5)$.   □

Remark 1 Though R_4 is constructed differently, can we expect similar result for $R_k, k > 4$? This is answered affirmatively by Theorem 6.1.

2   What can we say about the maximum size of R_4? Or how many of the pedigrees in P_5 can be mutually adjacent? For this, we need to look at the adjacency structure of the $1 - skeleton$ or the graph of $conv(P_5)$.

   However, it is helpful to present the nonadjacency among pedigrees and then conclude on the mutual adjacency of pedigrees. In Fig. 6.1 we have nonadjacent pedigrees connected by a coloured arrow. One of a pair of nodes connected by each coloured arrow forms a set of mutually adjacent pedigrees. For instance, Pedigrees printed in red are mutually adjacent. We can see that the maximum size of R_4 cannot be more than 6. In general, we need an upper bound on $|R_k|$. This is done in Theorem 6.2.

Theorem 6.1 *Pedigrees in R_{k-1} are mutually adjacent in $conv(P_k)$. That is, the corresponding characteristic vectors of any two rigid pedigrees, are adjacent in the $1 - skeleton$ or graph of $conv(P_k)$.*   ♡

Proof Suppose $P^{[i]} \in R_{k-1}, i = 1, 2$ are nonadjacent in $conv(P_k)$. Recall the construction of R_{k-1} from Sect. 5.3.2. Any pedigree $P^{[i]}$ that is included in R_{k-1} is such that it is constructed by either [i] a unique pedigree path in N_{k-2} or [ii] a unique rigid pedigree from R_{k-2} extended by $L_i = (e_{k-1}^{[i]}, e_k^{[i]})$ that corresponds to a rigid arc in F_{k-1}. We have a fixed positive weight $\mu(P^{[i]})$ attached to $P^{[i]}$, making it a rigid pedigree.

---

[1] Strictly speaking we should be saying the indicator/characteristic vector of the pedigree, since the polytope is convex hull of such vectors.

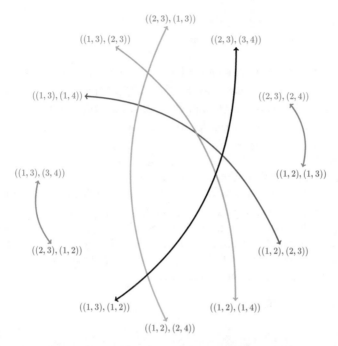

Fig. 6.1 Nonadjacency structure of $conv(P_5)$. Any coloured arrow shows a pair of nonadjacent pedigrees

Let $P^{[i]} = (e_4^{[i]}, \ldots, e_{k-1}^{[i]}, e_k^{[i]})$, $i = 1, 2$. Therefore, $P'^{[i]} = (e_4^{[i]}, \ldots, e_{k-1}^{[i]})$ is the unique pedigree path ending in $e_{k-1}^{[i]}$ for the link, $L_i = (e_{k-1}^{[i]}, e_k^{[i]})$ that is rigid.

We prove the theorem by considering different cases.

Case 1:
$P'^{[1]} = P'^{[2]}$

In this case, L_i must be different for different i, otherwise the pedigrees $P^{[i]}$, $i = 1, 2$, will not be different. As the two pedigrees differ in only in one component (the last component), applying Lemma 4.9 (Lemma 5.3 in [14]), we assert that $X^{[i]}$, $i = 1, 2$, are adjacent in $conv(P_k)$. Hence the result.

Case 2:
$P'^{[1]}$ *and* $P'^{[2]}$ *are different* Since $conv(P_k)$ is a combinatorial polytope as seen in Sect. 4.6.1 [11], if $X^{[1]}$ and $X^{[2]}$ are non-adjacent, there exist $X^{[3]}$ and $X^{[4]}$ such that $\frac{1}{2}(X^{[1]} + X^{[2]}) = \frac{1}{2}(X^{[3]} + X^{[4]})$. And the support for $X^{[3]}$ and $X^{[4]}$ is the same as that of $X^{[1]}$ and $X^{[2]}$. Therefore, a flow ϵ can be rerouted through the paths corresponding to $X^{[3]}$ and $X^{[4]}$, where $0 < \epsilon \leq$ minimum of $\{\mu(X^{[i]}), i = 1, 2\}$.

Sub-case a: $e_{k-1}^{[1]} = e_{k-1}^{[2]}$

In this case, we should have $e_k^{[1]}$ not equal to $e_k^{[2]}$. And $P^{[3]}$ or $P^{[4]}$ should end in $e_k^{[1]}$ or $e_k^{[2]}$. Without loss of generality, let $P^{[3]}$ end in $e_k^{[1]}$. $P^{[3]}$ is different from $P^{[1]}$

but corresponds to the same rigid link, $(e_{k-1}^{[1]}, e_k^{[1]})$, this contradicts the uniqueness of the path for this link. Hence the result.

Sub-case b: $e_k^{[1]} = e_k^{[2]}$

[The proof is similar to Sub-case a.] In this case, we should have $e_{k-1}^{[1]}$ not equal to $e_{k-1}^{[2]}$. And $P^{[3]}$ or $P^{[4]}$ should have $e_{k-1}^{[1]}$ or $e_{k-1}^{[2]}$. Without loss of generality, let $P^{[3]}$ have $e_{k-1}^{[1]}$. $P^{[3]}$ is different from $P^{[1]}$ but corresponds to the same rigid link, $(e_{k-1}^{[1]}, e_k^{[1]})$, this contradicts the uniqueness of the path for this link. Hence the result.

Sub-case c: $e_{k-1}^{[1]}$ not equal to $e_{k-1}^{[2]}$ & $e_k^{[1]}$ not equal to $e_k^{[2]}$

If either $P^{[3]}$ or $P^{[4]}$ ends with L_1 or L_2, then this contradicts the uniqueness of the respective path for these links. Hence the result.

If one of $P^{[3]}$ or $P^{[4]}$ ends with $(e_{k-1}^{[1]}, e_k^{[2]})$, then the other should end with $(e_{k-1}^{[2]}, e_k^{[1]})$. Then consider rerouting a flow of $\epsilon > 0$ along the paths $P^{[i]}, 1 = 1, \ldots, 4$, by increasing the flow along $P^{[3]}$ and $P^{[4]}$ and reducing the same along the paths corresponding to $P^{[1]}$ and $P^{[2]}$. Thus, the links $L_i, i = 1, 2$ are not rigid. This contradicts the fact that links are rigid, as $P^{[i]} \in R_k, i = 1, 2$. Hence the result.

This completes the proof of the theorem.   □

Next, we discuss the size of R_{k-1}.

Theorem 6.2 *Given $k \geq 5$ and (N_{k-1}, R_{k-1}, μ), is well-defined, we have* $|R_{k-1}| \leq dim(\Lambda_k(X)) + 1$, *where* $\Lambda_k(X) = \{\lambda : \sum_{i \in I(\lambda)} \lambda_i P_i = X/k,$ $\sum_{i \in I(\lambda)} \lambda_i = 1, \lambda_i > 0\}$. *(Same as defined in Chap. 4, Definition 4.5.)* ♡

Proof (N_{k-1}, R_{k-1}, μ), is well-defined, implies $X/k \in conv(P_k)$. However, the dimension of $\Lambda_k(X)$, depends on whether X is a vertex or it is on an edge or is in any other face of $conv(P_k)$. From Theorem 6.1, we know that the pedigrees in R_{k-1} are mutually adjacent. Hence they form a simplex. Therefore there can be at most $dim(\Lambda_k(X)) + 1$ pedigrees in R_{k-1}. Hence the result.   □

Corollary 6.1 *Given $k \geq 5$ and (N_{k-1}, R_{k-1}, μ), is well-defined, we have* $|R_{k-1}| \leq \tau_k - k + 4$.   ♡

Proof From Theorem 6.2, $|R_{k-1}| \leq dim(\Lambda_k(X)) + 1$. But $dim(\Lambda_k(X))$ can be at most equal to the dimension of $conv(P_k)$, which is $\tau_k - (k - 3)$ (see for a proof given in Chap. 7, Sect. 7.5 [12]).   □

The above results are useful in estimating the computational complexity of the proposed approach to check the membership of a given X in the pedigree polytope.

6.1.2 *Estimating the Computational Burden at Different Steps of the Framework*

In this subsection, I will go into estimating the upper bounds on the computational burden at each of the steps of the algorithmic framework given by Fig. 6.2. Although there may exist better estimates on the worst-case complexity in many of the problems encountered, I do not go into such discussions at this stage.

Step:1a Producing a certificate for $X \in P_{MI}(n)$: This involves [1] checking $X \geq 0$, [2] checking $\sum_{e \in E_{k-1}} x_k(e) = 1$ for $4 \leq k \leq n$, [3] checking sequentially $U^{l+1} \geq 0, 3 \leq l \leq n - 1$, where U is as defined in the reformulated Problem of $MI - relaxation$. This can be done in time strongly polynomial in n, involving only additions and subtractions.

FRAMEWORK: *Membership Checking Steps and Procedures*
Purpose: Given the number of cities, $n \geq 5$, the framework outlines the stages and conditions to be met for membership of $X \in Q^{\tau_n}$ in pedigree polytope $conv(P_n)$.
   1. It requires a certificate that X belongs to $P_{MI}(n)$,
   2. It requires a certificate for F_4 feasibility,
   3. For each $5 \leq k < n$, it sequentially checks for a certificate for F_k feasibility, before it checks for a certificate for $X/k + 1 \in conv(P_{k+1})$,
   4. **until finally a certificate for $X \in conv(P_n)$ is produced.**
   5. Otherwise, $X \notin conv(P_n)$ is established when we fail to produce any of the earlier mentioned certificates.
Step:1a If $X \in P_{MI}(n)$ Proceed further, otherwise **Stop.**
Output: Evidence exists for $X \notin conv(P_n)$, $X \notin P_{MI}(n)$.

Step:1b If F_4, is feasible, identify, all $P \in R_4$ using FFF algorithm. Set $k = 5$, proceed further, otherwise **Stop.**
Output: Evidence exists for $X \notin conv(P_n)$, F_4 is not feasible or $X/5 \notin conv(P_5)$.

Step:2a Find capacity $C(L)$ of each link, $L \in V_{[k-3]} \times V_{[k-2]}$, that is, the maximum flow in the restricted network, $N_{k-1}(L)$. Construct F_k.
Step:2b If F_k, is feasible, proceed further, otherwise **Stop.**
Output: Evidence exists for $X \notin conv(P_n)$, F_k is not feasible.

Step:3 Use FFF procedure and identify rigid and dummy arcs in F_k.
Find rigid paths $P \in R_k$ along with respective rigid flow, μ_P.

Update capacities of nodes and arcs in N_k. Construct (N_k, R_k, μ).
Step:4 Construct and solve the multicommodity flow problem, *Problem* 5.1.
Step:5 Set $k = k + 1$. If the optimum total flow, z^* is equal to the maximum possible, that is, z_{max}, then, $X/k \in conv(P_k)$ and proceed further, otherwise **Stop.**
Output: Evidence exists for $X \notin conv(P_n)$, $X/k \notin conv(P_k)$.

Step:6 If $k < n$, **Repeat Step:2a** on wards, otherwise **Stop.**
Output: $X \in conv(P_n)$.

Fig. 6.2 Algorithmic framework for checking membership in pedigree polytope

Step:1b Constructing and checking F_4, feasibility. Solving F_4 is equivalent to solving a *FAT* problem, with maximum 3 origins and 6 destinations with arcs connecting $[4 : e]$ with $[5 : e']$, if e is a generator of e'. And applying the *FFF* algorithm for this case is equally easy.

Step:2a Find capacity of each link, L, that is, the maximum flow in the restricted network, $N_{k-1}(L)$. Construct F_k.

Firstly, constructing the restricted network $N_{k-1}(L)$ involves applying the deletion rules, Definition 5.2. Each of these rules is either [i] deleting a certain set of nodes or [ii] checking whether for a node all its generators have been deleted. These can be done in time linear in the number of nodes in the network. And [ii] may be repeated at most $|\mathcal{V}(N_{k-1})|$ times, for each node in the network. Finally, finding the subnetwork induced by the remaining nodes, $\mathcal{V}(N_{k-1}(L))$, can also be done efficiently, to obtain $N_{k-1}(L)$.

Secondly, we need to estimate the complexity of finding the capacity for each link, L. Each of these problems can be solved in time polynomial in the number of nodes and arcs in the network $N_{k-1}(L)$ (see for instance [2]). This is where the results proved in Sect. 6.1.1 become relevant. From Theorem 6.2 and Corollary 6.1, we have a polynomial bound on the cardinality of R_{k-1}, namely, $|R_{k-1}| \leq \tau_k - k + 4$. In N_{k-1} in the last layer we have a maximum of p_{k-1} nodes. In each of the other layers, we have $\leq p_l + \tau_l$ nodes in layer $l > 4$. Therefore, at most $\sum_{l=5}^{k-1}(p_{l-1} + \tau_l) + p_3 + p_{k-1} = \sum_{l=5}^{k} \tau_l$ nodes are possible. Therefore, the number of nodes is $\leq (k-5) \times \tau_k$. The number of arcs in N_{k-1} can be at most $\sum_{l=5}^{k-1}(p_{l-1} + \tau_l)p_l + 3 \times 6$. The number of links, L we have are at most $p_{k-1} \times p_k$ ($< k^4$), as we consider links between nodes in $V_{[k-3]}$ and $V_{[k-2]}$. Thus, both the number of nodes and arcs in the network are bounded above by a polynomial in k for each link L. Or each of these flow problems can be solved in time strongly polynomial[2] in k.

Step:2b Checking F_k, feasibility. Solving a *FAT* problem can also be done in time polynomial in $|\mathcal{O}|$, $|\mathcal{D}|$ and $|F|$, where F is the set of forbidden arcs [155]. Construction of $F_k, k > 4$ involves nodes corresponding to pedigrees in R_{k-1} as origins, apart from the nodes in $V_{[k-3]}$. Therefore it is crucial to ensure the number of origins in F_k does not grow to be exponentially large.

As seen earlier, we have a polynomial bound on the cardinality of R_{k-1} namely, $|R_{k-1}| \leq \tau_k - k + 4$.

Thus the number of origins in F_k will never grow exponentially.

Step:3 Use *FFF* procedure and identify rigid and dummy arcs in F_k.

Frozen Flow Finding (*FFF*) algorithm to identify rigid and dummy arcs, as noted earlier in Chap. 2 can be done in linear time in the size of the graph G_f corresponding to the problem F_k. The size of G_f is at most $(p_{k-1} + |R_{k-1}|) \times p_k + p_{k-1} + |R_{k-1}| + p_k$.

Find rigid paths for $P \in R_k$ along with respective rigid flow μ_P. Rigid paths in R_k either come from a unique path in N_{k-1} extended by a rigid arc or from a rigid path in R_{k-1} extended by a rigid arc. We store the unique path $P_{unique}(L)$ corresponding a link L, in case it was obtained while finding the max flow in $N_{k-1}(L)$.

Thus (N_k, R_k, μ) can be constructed in time polynomial in k.

---

[2] Currently almost in linear time in the number of edges, one can solve a network max-flow problem. https://doi.org/10.48550/arXiv.2203.00671.

Step:4 The next task is to find whether a feasible multicommodity flow exists with an objective value equal to z_{max}. Since we have only a *linear and not integer* multicommodity problem to be solved, we need to solve only a linear programming problem. This linear programming problem is a combinatorial linear programming as defined in Chap. 1. This problem can be solved using Tardos's [159] algorithm in strongly polynomial time in the input size of the dimension of the Problem 5.1. And the input size is polynomial in k. And we have to solve at most $n - 4$ such problems in all. Thus, we can check using the Framework Fig. 6.2 in strongly polynomial time whether $X \in conv(P_n)$.

Here I have not gone for tight bounds for the computational requirements, as the purpose is to prove a crucial result that the necessary and sufficient condition given by Theorem 5.5 can be checked efficiently and is stated as Theorem 6.3.

Theorem 6.3 *Given n, $X \in P_{MI}(n)$, and $X/n - 1 \in conv(P_{n-1})$. Then checking whether there exists a solution for the multicommodity flow problem (Problem 5.1 with $k = n$), with $z^* = \sum_{s \in S_k} v^s = z_{max}$, can be done efficiently, that is, it is strongly polynomial in n.* ♡

6.2 Concluding Remarks

What remains is checking some technicalities before I can claim the unexpected consequences of Theorem 6.3. Chapter 7 is devoted to the task of explaining those technicalities and checking they are met.

6.3 Appendix: Illustrative Example

I run through an example clarifying the construction of the network and other related computations using an example from [7]. We are given $n = 10$, and a $X \in R^{\tau_n}$ for checking its membership in $conv(P_n)$.

Example 6.1 Consider X, corresponding to a problem of size 10, with the following x_{ijk} values[3]:

$$x_{124} = 1, \quad x_{135} = \tfrac{3}{4}, \quad x_{145} = \tfrac{1}{4}, \quad x_{246} = \tfrac{2}{4}, \quad x_{356} = \tfrac{1}{4},$$

$$x_{456} = \tfrac{1}{4}, \quad x_{237} = \tfrac{1}{4}, \quad x_{157} = \tfrac{1}{4}, \quad x_{467} = \tfrac{1}{4}, \quad x_{567} = \tfrac{1}{4},$$

$$x_{238} = \tfrac{1}{4}, \quad x_{148} = \tfrac{2}{4}, \quad x_{468} = \tfrac{1}{4}, \quad x_{269} = \tfrac{2}{4}, \quad x_{189} = \tfrac{1}{4},$$

$$x_{689} = \tfrac{1}{4}, \quad x_{3510} = \tfrac{2}{4}, \quad x_{4710} = \tfrac{1}{4}, \quad x_{6710} = \tfrac{1}{4}.$$

---

[3] This is the example given in [7] to show the necessary condition given in [12] is not sufficient.

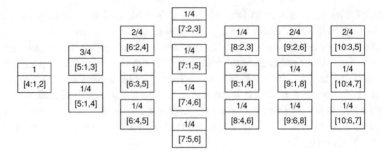

Fig. 6.3 Capacities of the nodes for Example 6.1, $X \in P_{MI}(10)$

Fig. 6.4 Networks
corresponding to $k = 4$ & 5.
In both cases R_k is nonempty
and N_k is empty

3/4		3/4
[4:1,2]		[5:1,3]

1/4		1/4
[4:1,2]		[5:1,4]

[a] Network for *k*=4

2/4		2/4		2/4
[4:1,2]		[5:1,3]		[6:2,4]

1/4		1/4		1/4
[4:1,2]		[5:1,3]		[6:3,5]

1/4		1/4		1/4
[4:1,2]		[5:1,4]		[6:4,5]

[b] Network for *k*=5

Firstly, it can be checked that $X \in P_{MI}(10)$, as [1] X is nonnegative and the
equality constraints of $P_{MI}(10)$ are satisfied for each k, then [2] by substituting X
in the inequality constraints for each (i, j), we find all are satisfied. Figure 6.3 gives
the positive $x_k(e)$ values, as capacities of the nodes along with the node description
$[k : e]$.

In F_4, we have $V_{[1]} = \{[4 : 1, 2]\}$ and $V_{[2]} = \{[5 : 1, 3], [5 : 1, 4]\}$. Since $[4 : 1, 2]$
is a generator for all the nodes in $V_{[2]}$, the set of arcs is $\mathcal{A}(F_4) = \{([4 : 1, 2],$
$[5 : 1, 3]), ([4 : 1, 2], [5 : 1, 4])\}$. Here, both arcs are rigid and so R_4 contains them,
and N_4 is empty. In Fig. 6.4a, we have R_4. In this figure, we have the capacities
of the nodes along with node description, bold lines indicate rigid pedigrees. Next,
F_5 is checked feasible, and the resulting R_5 has three rigid pedigrees as shown in
Fig. 6.4b. Restricted networks for $k = 6$ are then constructed. Firstly, notice that
for each rigid pedigrees in R_5 we can have an arc connecting that pedigree to a
node, $v = [7 : e_\beta] \in V_{[4]}$, with $x(v) > 0$, such that a generator of e_β is in the rigid
pedigree. For example, $[7 : 2, 3]$ is connected to all the three pedigrees in R_5, as
$[4 : 1, 2]$ is a generator of $(2, 3)$, with capacity 1/4 and is present in all the three

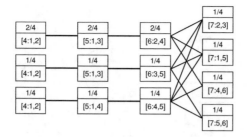

Fig. 6.5 Network corresponding to $k = 6$, R_6 is empty

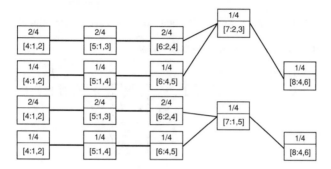

Fig. 6.6 Restricted networks for $([7 : e], [8 : 4, 6])$ for $e = (2, 3)$ & $(1, 5)$

Fig. 6.7 Network for F_7

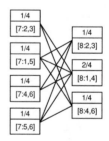

pedigrees. But $[7 : 4, 6]$ is connected to only two of them as the third, namely, $([4 : 1, 2], [5 : 1, 3], [6 : 3, 5])$ does not contain any generator of $(4, 6)$. Thus, we obtain the arc set for F_6. The node set has three origins corresponding to the rigid pedigrees in R_5 and the sink nodes are $[7 : 2, 3]$, $[7 : 1, 5]$, $[7 : 4, 6]$, and $[7 : 5, 6]$. Figure 6.5 gives F_6 which is feasible, and no arc is rigid. So $R_6 = \emptyset$, $\mu = 0$. We now observe that as in Remark (2) after Theorem 5.5, F_6 corresponds to a $FAT_6(\lambda)$ problem and so $X/7 \in conv(P_7)$. Thus, N_6 is same as F_6. (N_6, R_6, μ) is well-defined.

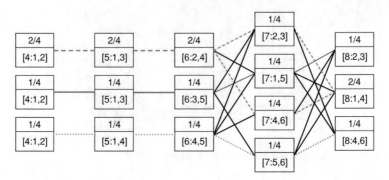

Fig. 6.8 Network for $k = 7$, N_7. Commodity flows are shown in green, red and blue

Restricted network for $L = ([8:2,3], [9:2,6])$ with $C(L) = 1/4$:

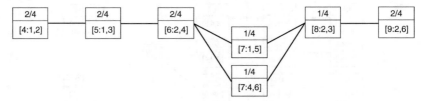

Restricted network for $L = ([8:1,4], [9:2,6])$ with $C(L) = 2/4$:

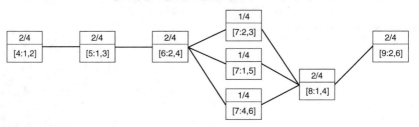

Restricted network for $L = ([8:1,4], [9:1,8])$ with $C(L) = 2/4$:

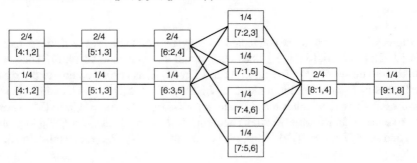

Fig. 6.9 Restricted networks $N_7(L)$

Restricted network for _L = ([8:4,6], [9:6,8])_ with _C(L) = 1/4_:

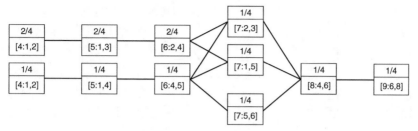

Restricted network for _L = ([8:4,6], [9:2,6])_ with _C(L) = 1/4_:

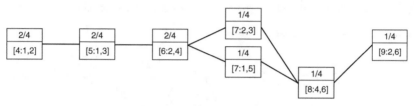

Fig. 6.10 Restricted networks $N_7(L)$ (continued)

Fig. 6.11 Network F_8

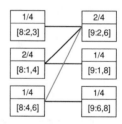

Next, F_7 is constructed using nodes in $V_{[4]}$ as origins and nodes in $V_{[5]}$ as sinks. The restricted networks for the links are considered for calculating the capacities of the arcs. Link ([7 : 2, 3], [8 : 2, 3]) is forbidden, we do not have this arc in F_7. Since both (2, 3)&(1, 4) has generators in [4 : 1, 2] and maximum flow possible into any node in $V_{[4]}$ is 1/4, we have all the arcs connecting $V_{[4]}$ and [8 : 2, 3] and [8 : 1, 4], except for link ([7 : 2, 3], [8 : 2, 3]), which is forbidden. The link ([7 : 4, 6], [8 : 4, 6]) is forbidden as well. Restricted networks for other links ([7 : e], [8 : 4, 6]) are shown in Fig. 6.6 for $e = (2, 3)$&$(1, 5)$. The restricted network for ([7 : 5, 6], [8 : 4, 6]) has the unique path (([4 : 1, 2], [5 : 1, 4], [6 : 4, 5]), [7 : 5, 6]) having a maximum flow of 1/4. Thus, we have constructed F_7 as shown in Fig. 6.7 with all arc capacities 1/4, and F_7 is feasible with no rigid arc. Network N_7 is constructed as shown in Fig. 6.8. The multicommodity flow problem is feasible with $z_{max} = 1$. The pedigree paths shown in different colours (red, green and blue) with flows (1/4, 1/2 = (1/4 + 1/4), and 1/4) respectively show that $X/8 \in conv(P_8)$. Black arcs have zero flow. Next, we consider restricted networks for the links between $V_{[5]}$&$V_{[6]}$, as shown in Figs. 6.9 and 6.10. F_8 is feasible. *FFF* algorithm finds $L = ([8 : 4, 6], [9 : 2, 6])$ to be a dummy arc (coloured red), all other arcs are rigid with capacity 1/4 (Fig. 6.11).

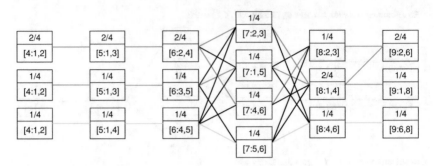

Fig. 6.12 Network N_8. Four pedigree paths are shown in blue, green and yellow

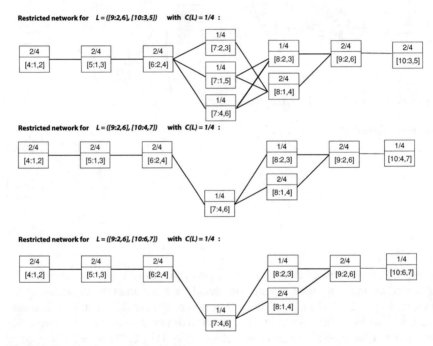

Fig. 6.13 Restricted networks for $N_8(L)$

N_8 as shown in Fig. 6.12 is constructed and the multi-commodity flow problem yields $z_{max} = 1$. The pedigree paths (blue, green, and yellow) with respective flows (1/2 = 1/4 +1/4), 1/4, and 1/4 are shown. We then consider Restricted Networks for the links between the last two stages (Figs. 6.13, 6.14 and 6.15).

Restricted network for $L = ([9{:}1{,}8], [10{:}3{,}5])$ **with** $C(L) = 1/4$:

Restricted network for $L = ([9{:}1{,}8], [10{:}4{,}7])$ **with** $C(L) = 1/4$:

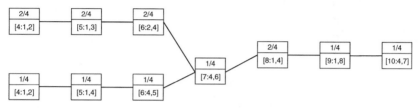

Restricted network for $L = ([9{:}1{,}8], [10{:}6{,}7])$ **with** $C(L) = 1/4$:

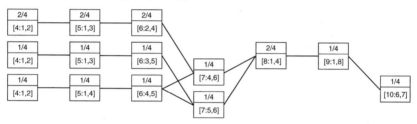

Fig. 6.14 Restricted networks for $N_8(L)$ (continued)

F_9 problem has all arc capacities 1/4, except for arc $([9{:}2, 6], [10{:}3, 5])$ which has capacity 2/4, however, there is no arc $([9{:}6, 8], [10{:}4, 7])$. The problem is feasible, and no arc is rigid. Network for F_9 is as shown in Fig. 6.16.

Next, we need to check that N_9 is well-defined, by solving the multicommodity flow problem for N_9. We see that the maximal flow for the multicommodity flow problem for N_9 has a maximum total flow of $7/8 < z_{max} = 1$. This implies $X \notin conv(P_{10})$. See Fig. 6.17 for the maximal multicommodity flow paths in $N9$.

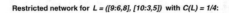

Restricted network for L = ([9:6,8], [10:3,5]) with C(L) = 1/4:

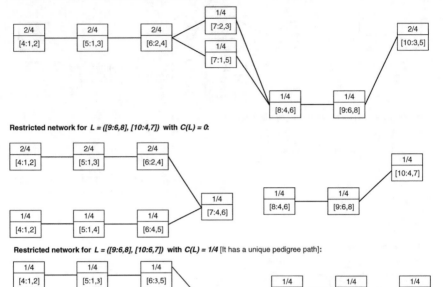

Restricted network for L = ([9:6,8], [10:4,7]) with C(L) = 0:

Restricted network for L = ([9:6,8], [10:6,7]) with C(L) = 1/4 [It has a unique pedigree path]:

Fig. 6.15   Restricted networks for $N_8(L)$ (continued)

Fig. 6.16   Network F_9

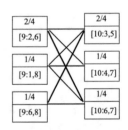

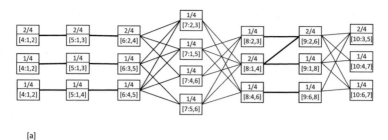

[a]

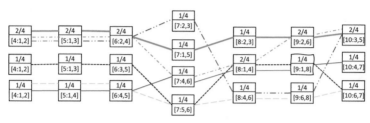

[b].

> **Legend:**
> in part [a], rigid paths and arcs are shown in bold.
> In part [b], weight of an arc is proportional to the flow value, here 2/8 or 1/8 respectively. Different
> colours represent different commodity flows with maximum multicommodity flow, 7/8. Five different
> paths are depicted using different line dashes.

Fig. 6.17   a Network for N_9 **b** Multicommodity flow paths

Notice that the capacity of the rigid path ([4 : 1, 2], [5 : 1, 3], [6 : 3, 5]) is 1/4 and only 1/8 flows out of this.

Chapter 7
Efficient Checking of Membership in Pedigree Polytope and Its Implications

Proof of NP = P

7.1 Introduction

Theoretical computer science, among other things, deals with the design and analysis of algorithms. When one wants to solve a problem efficiently using an algorithm, the amount of storage space and computational time required are considered and compared. The class of problems solvable in polynomial time by a Turing machine is designated as the class P (see e.g. [76, 108].) Another class of problems is the class NP, which consists of a problem that can be solved by a nondeterministic Turing machine in polynomial time. Polynomial-time algorithms received their prominence against slow exponential time approaches, often bordering on exhaustion of all solutions by brute force. As we noticed in Chap. 1 Jack Edmonds [64] called them *good algorithms*, defining them as "For an algorithm to be good we mean that there is a polynomial function $f(n)$ which for every n is an upper bound on the "amount of work" the algorithm does for any input of "size" n." [65], and presented a good algorithm for the matching problem, which has a linear programming formulation involving exponentially many constraints. We saw in Chap. 1 the seminal work by Stephen Cook [49] and the immediate recognition by Richard Karp [101] of its importance to combinatorial optimisation and integer programming, the so-called NP-complete problems, as opposed to polynomially solvable problems, received renewed attention. The problem of whether P = NP is one of the most outstanding problems in mathematics.

In this chapter, we discuss the complexity of checking the membership of a given X in the pedigree polytope. We will use the concept from the book, *Geometric algorithms* [82] by Grötschel, Lovász, and Schrijver to use our membership checking framework as a subroutine/oracle in finding a separating facet of $conv(P_n)$, that cuts off X if it is not in the pedigree polytope. In [82] the construction of Yudin and Nemirovskii [166] is used to establish conditions for the existence of a polynomial separation algorithm for a bounded convex body. The approach in [82], uses Leonid Khachiyan's ellipsoid algorithm twice. Given a polytope P, a, in the interior of P and

© The Author(s), under exclusive license to Springer Nature Singapore Pte Ltd. 2023
T. S. Arthanari, *Pedigree Polytopes*,
https://doi.org/10.1007/978-981-19-9952-9_7

a $x \notin P$, recently, Jean François Maurras [121] has given under certain conditions, a simple construction to identify a violated facet of the polytope, P, whose supporting hyperplane separates x from P. This uses a polynomial number of calls to an oracle checking membership in P. We consider an alternative polytope associated with *STSP* and verify whether Maurras's construction is possible for this polytope. This is what led us to study the membership problem for the pedigree polytope. Chapters 4, 5, and 6 introduced the concepts required to study an alternative polytope $conv(A_n)$ associated with *STSP*. A_n is defined in this chapter, and Sect. 7.2 checks the conditions for the existence of a polynomial separation algorithm for $conv(A_n)$. Thus showing the membership problem of the pedigree polytope can be solved efficiently leads to the unexpected theorem, $\mathsf{P} = \mathsf{NP}$.

7.2   Polytopes and Efficiency

Consider $P \subset R^d$, a *$\mathcal{H}$-polyhedron*, that is, P is the intersection of finitely many half-spaces, $\mathbf{a}_i X \leq a_0$, for $(\mathbf{a}_i, a_0) \in R^{d+1}$, for $i = 1, \ldots, s$. Recall [a] a bounded $\mathcal{H}$-polyhedron is indeed a $\mathcal{V}$-polytope, [b] the *affine rank* of a polytope P (denoted by $arank(P)$) is defined as the maximum number of affinely independent vectors in P, and [c] the *dimension* of a polytope P is denoted by $dim(P)$ and defined to be $arank(P)$ minus 1.

Let $P \subset R^d$ be a polytope. The *barycentre* of P is defined as

$$\bar{X} = 1/p \sum_{X^i \in vert(P)} X^i,$$

where p is the cardinality of $vert(P)$, the vertex set of P. [See Chap. 2, or [168] for an introduction to polytopes.]

Given $n \in Z$ the *input size* of n is the number of digits in the binary expansion of the number n plus 1 for the sign, if n is non-zero. We write,

$$\langle n \rangle = 1 + \lceil \log_2 (n + 1) \rceil.$$

Input size of n, $\langle n \rangle$, is also known as the *digital size* of n.

Given $r = p/q$ a rational number, where p and q are mutually prime, that is, $gcd(p, q) = 1$, we have input size of r given by

$$\langle r \rangle = \langle p \rangle + \langle q \rangle.$$

For every rational r we have $|r| \leq 2^{\langle r \rangle - 1} - 1$.

Definition 7.1 (*Rationality Guarantee*) Let $P \subset R^d$ be a polytope and ϕ and ν positive integers. We say that P has *facet complexity* at most ϕ if P can be described as the solution set of a system of linear inequalities each of which has input size $\leq \phi$. We say, P has vertex complexity at most ν if P can be written as $P = conv(V)$, where $V \subset Q^d$ is finite and each vector in V has input size $\leq \nu$. ♣

We have Lemma 7.1 from [82] connecting facet and vertex complexities.

Lemma 7.1 *Let $P \subset R^d$ be a non empty, full dimensional polytope. If P has vertex complexity at most ν, then P has facet complexity at most $3d^2\nu$.* ♡

7.3 Membership and Optimisation

Imagine you are at the gate of an exclusive Internet Club. If you need to be accepted into the club, you have to produce a password/membership card. The password is checked for validity and then you are admitted or not, accordingly. A protocol-checking security program does the verification at the gate. Some clubs have such checking at more than one gate. Similar is the case with an exclusive set of interest, like a polytope, $P \subset Q^d$. If any $y \in Q^d$ is presented with its encoding as a password, we need an algorithm that checks the validity of membership of y in P. This problem is addressed next.

Problem 7.1 (*Membership Problem*) Given a polytope $P \subset Q^d$, and $Y \in Q^d$, the problem of deciding whether $Y \in P$ or not, is called the membership problem for/of P.

Definition 7.2 (*Quick Protocol*) Given a polytope $P \subset Q^d$, with facet complexity at most ϕ. Let MemAl(P, Y, Answer) be an algorithm to solve the membership problem, where P is known to MemAl not necessarily explicitly, and on the input of $Y \in Q^d$ having input size $\langle Y \rangle$, MemAl halts with Answer = yes, if $Y \in P$ and Answer = no, otherwise. If the membership checking time of MemAl is polynomially bounded above by a function of $(d, \phi, \langle Y \rangle)$ we say MemAl is a quick protocol or a good algorithm. ♣

Problem 7.2 (*Separation Problem*) Given a polytope $P \subset Q^d$, and a $Y \in Q^d$, the problem to decide whether $Y \in P$, and if $Y \notin P$ then identifying a hyperplane that separates P and Y is called the separation problem for P. Identifying a hyperplane is achieved through finding a vector $a \in Q^d$ such that $aX < aY \ \forall \ X \in P$.

Theorem 7.1 (Yudin and Nemirovskii [82, 166]) *Let $X_0 \in \mathrm{int}(P)$ be given. $\exists$ an algorithm that solves the separation problem for the polytope $P \subset Q^d$, in time polynomially bounded by d, ϕ, $\langle X_0 \rangle$, and the running time of the membership algorithm* MemAl. ♡

For proof see *Geometric Algorithms* by Grötschel, Lovász and Shrijver. The translations in English from the original Russian articles by Yudin and Nemirovskii, have been further paraphrased in the above book, without which trying to dig out this result from the originals would have been difficult. The authors of the survey paper on the ellipsoidal method[1] note this cryptic description of the shallow-cut ellipsoidal method, and other results appearing therein. Section 4.3 of Geometric algorithms deals with the above theorem.

Given an LP problem, we assume we know all defining equalities and inequalities. However, we can state the underlying polytopal optimisation problem formally as follows:

Problem 7.3 (*Linear Optimisation Problem*) Given a non empty polytope $P \subset Q^d$, and a $C \in Q^d$, the problem of finding a $X^* \in P \ni CX^* \leq CX \ \forall \ X \in P$ is called the linear optimisation problem for P.

An important result that we need from *Geometric Algorithms* on linear optimisation problem is stated below:

Theorem 7.2 (Grötschel, Lovász and Shrijver [82]) *Given the input (d, ϕ, SEP_{Al}, C), where d and ϕ are natural numbers, and a separation algorithm, SEP_{Al}, for some rational nonempty polytope $P \subset Q^d$, with facet complexity ϕ, and $C \in Q^d$, then $\exists$ an algorithm that solves the linear optimisation problem for P in time polynomially bounded by d, ϕ, $\langle C \rangle$, and the running time of the separation algorithm.* ♡

(For proof see [82] or later books on LP or combinatorial optimisation cited in the book.)

---

[1] By R. G. Bland, D. Goldfarb and M. J. Todd, Operations research, Vol. 29, pp. 1039–1091 (1981).

Combining the two theorems ([Yudin and Nemirovskii] & [Grötchel, Lovász, and Shrijver]), we have the very important implication given below:

> **Theorem 7.3** (YuNGLS Theorem) *Given a polytope $P \subset Q^d$ such that,*
> *1.* $\dim(P) = d$,
> *2. P is rationality guaranteed (facet complexity $\leq \phi$),*
> *3. An $X_0 \in \text{int}(P)$ is given,*
> *4. A quick protocol is available to check membership in P,*
> *then $\exists$ an algorithm to solve the linear optimisation problem over P for a given*
> *$C \in Q^d$ in time polynomially bounded by d, ϕ, $\langle X_0 \rangle$ & $\langle C \rangle$.* ♡

This result, its converse and a few other equivalent results are summarised in Lovász [115] as Theorem 2.3.3. The term "oracle" there can be read equivalently as "subroutine." Here, only the required specialised results for full-dimensional, non-empty polytopes are stated, although more general ones are available. [See [82, 115, 152] for such general results for polyhedrons and convex bodies.]

Maurras [121] shows that an intuitively appealing construction is possible for the separation problem of a polytope, by finding a hyperplane separating the polytope and a point not in the polytope, after a polynomial number of calls to a membership oracle. The conditions under which this is possible are the same as that of [166], namely,

Assumption (Maurras's Conditions)

1. The polytope P is well defined in the d-dimensional space of Q^d of rational vectors. (There is a bound on the encoding length of any vertex of P. The polytope is rationality guaranteed.)
2. P has non-empty interior.
3. $a \in int(P)$ is given.

Grötschel, Lovász, and Schrijver in [82] use a construction due to Yudin and Nemirovskii [166] to devise a polynomial algorithm for finding a separating plane using a membership oracle, when a convex set K instead of the polytope P is considered. However, this algorithm requires in addition the radii of the inscribed and circumscribed balls, it also uses twice the ellipsoid algorithm by Leonid Khachiyan.

Maurras' polynomial construction and its justification given in [121], for separation using membership oracle for well defined non empty polyhedra, 'can be considered to lie in the folklore of mathematical programming', as demonstrated later by Maurras in [122]. In Sect. 7.5 we present an excerpt from [121].

Next, we check that the Assumptions of Theorem 7.3 are met for a polytope closely related to the *STSP* polytope.

7.3.1 Dimension of the Pedigree Polytope

Grötschel and Padberg [85] have shown that the dimension of the *STSP* polytope Q_n is $d_n = \frac{n(n-3)}{2}$. It is easy to see $dim(Q(n)) \leq d_n$ as every n-tour satisfies the 2-matching constraint. Their proof uses the graph theoretic lemma on the partitioning of the edge set of E_n into s Hamiltonian cycles and a perfect matching of the remaining edges in E_n, where $n = 2s + 2$ or $2s + 3$, $s \geq 1$ depending on n is even or odd respectively. Essentially, they show that a set of d_n linearly independent n-tours can be selected proving the dimension of $Q(n) \geq d_n$.

Here we show that the dimension of $conv(P_n)$ is $\tau_n - (n - 3)$, recall that τ_n is the number of coordinates of a $X \in P_n$.

Recall the following Eq. 4.15 from Chap. 4 which facilitates the subsequent discussion.

In general, let $E_{[n]}$ denote the matrix corresponding to Eq. 7.1:

$$x_k(E_{k-1}) = 1, \ k \in V_n \setminus V_3 \qquad (7.1)$$

Let $\mathcal{D} = \{X \in R^{\tau_n} \mid E_{[n]}X = \mathbf{1}_{n-3}\}$. So $dim(\mathcal{D}) = \tau_n - (n - 3)$. As every pedigree satisfies Eq. 7.1, $conv(P_n) \subset \mathcal{D}$, therefore

$$dim(conv(P_n)) \leq \tau_n - (n - 3) \stackrel{def}{=} \alpha_n.$$

Therefore, like Q_n, $conv(P_n)$ is not a full-dimensional polytope. It is possible to show the other way inequality by selecting α_n pedigrees in P_n that are linearly independent as in the proof of the dimension of Q_n [85].

We next direct our search for a full-dimensional polytope with dimension α_n. And it is obtained as a projection of the pedigree polytope.

Let $e^k = (k - 2, k - 1)$, and $E'_{k-1} = E_{k-1} \setminus \{e^k\}$, for $k \in V_n \setminus V_3$.

Lemma 7.2 *Given $k \in V_{n-3} \setminus V_3$, and $e = (i, j) \in E'_k$, we can select a pedigree in P_n such that $x_{k+1}(e) = 1, x_{k+2}((i, k + 1)) = 1$ and $x_l(e^l) = 1, k + 3 \leq l \leq n$.* ♡

Proof Given k and e select a k-tour containing e as an edge and insert $k + 1$ in e obtaining a $k + 1$-tour. Choose $(i, k + 1)$ for insertion of $k + 2$ obtaining a $k + 2$-tour, which has $e^{k+3} = (k + 1, k + 2)$ as an edge. Now extend this $k + 2$-tour to a n-tour by sequentially inserting l in e^l for l ranging from $k + 3$ through n. Consider the corresponding pedigree $X \in P_n$. This has the required property.   □

Definition 7.3 (*Projection M*) Given $X \in P_n$, consider the transformation $Y = MX$, where M deletes the p_{k-1}^{th} component of $\mathbf{x}_k$ in X, giving a vector $Y \in B^{\alpha_n}$. ♣

The projection M is given by the matrix

$$
M = \begin{bmatrix}
I_{p_3-1} & 0 & 0 & 0 & \cdots & 0 \\
0 & 0 & I_{p_4-1} & \vdots & & \vdots \\
\vdots & \vdots & & \cdots & 0 & 0 \\
0 & 0 & \cdots & & I_{p_{n-1}-1} & 0
\end{bmatrix}.
$$

Notice that Y is the compact string that has the information contained in X, but there is some redundancy in X, namely, for any k, the last component of $\mathbf{x}_k$ does not say anything more than what is already said in the $p_{k-1} - 1$ preceding components. This is so because given $X \in P_n$ for each k we have a unique edge $e \in E_{k-1}$ such that $x_k(e) = x_k(E_{k-1}) = 1 \ \forall \ k$. (We have, $x_k(e^k) = 1 - x_k(E'_{k-1})$).

Theorem 7.4 *Let* $A_n = \{Y \in B^{\alpha_n} \mid Y = MX, X \in P_n\}$. *The dimension of the polytope* $conv(A_n)$ *is* α_n. ♡

Proof Since $conv(A_n) \subset R^{\alpha_n}$, $dim(conv(A_n)) \leq \alpha_n$. Suppose $dim(conv(A_n)) < \alpha_n$. This implies that there exists a hyperplane $CY = c_0$ with $(C, c_0) \in R^{\alpha_n+1}$ such that every vertex of $conv(A_n)$ lies on this hyperplane. Let $C = (\mathbf{c}_4, \ldots, \mathbf{c}_n)$, where $\mathbf{c}_k$ denotes the component corresponding to $(c_k(e), e \in E'_{k-1})$.

Claim 1: $c_0 = 0$.
Proof of Claim 1: Consider the pedigree $(e^4, \ldots, e^n)$. Then the corresponding $Y \in A_n$ is $\mathbf{0}$. So $CY = 0$ which implies $c_0 = 0$ as required.
Claim 2: $\mathbf{c}_4 = \mathbf{0}$.
Proof of Claim 2: Consider Y corresponding to the pedigree given by

$$((1, 2), (1, 4), e^6, \ldots, e^n).$$

Now $CY = 0$ implies $c_4((1, 2)) + c_5((1, 4)) = 0$. But the pedigree $((1, 3), e^5, \ldots, e^n)$ yields $c_4((1, 3)) = 0$; and the pedigree $((1, 3), (1, 4), e^6, \ldots, e^n)$ yields $c_4((1, 3)) + c_5((1, 4)) = 0$. Together we have $c_5((1, 4)) = 0$, and so $c_4((1, 2)) = 0$. Hence $\mathbf{c}_4 = \mathbf{0}$.

This forms the basis for the proof by induction on k. Assuming $\mathbf{c}_4, \ldots, \mathbf{c}_k$ are all zero vectors we shall show that $\mathbf{c}_{k+1} = \mathbf{0}$.

Let $e = (i, j) \in E'_k$. From Lemma 7.2 we have a pedigree in P_n such that $x_{k+1}(e) = x_{k+2}((i, k+1)) = 1$ and $x_l(e^l) = 1, k+3 \le l \le n$. For the corresponding Y we then have $CY = 0 + c_{k+1}(e) + c_{k+2}((i, k+1)) + 0 = 0$. This implies

$$c_{k+1}(e) + c_{k+2}((i, k+1)) = 0. \tag{7.2}$$

Consider a k-tour that contains the edge (i, k). Let Q be the corresponding pedigree in P_k. From $Y \in A_n$ corresponding to the pedigree $(Q, (i, k), e^{k+2}, \ldots, e^n)$, we get

$$c_{k+1}((i, k)) = 0. \tag{7.3}$$

From $Y \in A_n$ corresponding to the pedigree $(Q, (i, k), (i, k+1), e^{k+3}, \ldots, e^n)$, we get
$$c_{k+1}((i, k)) + c_{k+2}((i, k+1)) = 0. \tag{7.4}$$

Now using Eqs. 7.3 and 7.4 we have

$$c_{k+2}((i, k+1)) = 0. \tag{7.5}$$

Equations 7.2 and 7.5 yield $c_{k+1}(e) = 0$. Thus we have $\mathbf{c}_{k+1} = \mathbf{0}$.
Hence $C = \mathbf{0}$ and $c_0 = 0$ implying $dim(conv(A_n)) = \alpha_n$. □

Now any $Y \in conv(A_n)$ can be extended to a vector in R^{T_n} by augmenting the last coordinate $y_k(e^k)$ to each component $\mathbf{y}_k$ of Y corresponding to $k \in V_n \setminus V_3$. It is easy to see, if $y_k(e^k) = 1 - y_k(E'_{k-1})$ for each k, then such a vector is in $conv(P_n)$. Therefore $dim(conv(P_n)) \ge \alpha_n$. Thus we have proved Theorem 7.5.

Theorem 7.5 *The dimension of the pedigree polytope $conv(P_n)$ is α_n.* ♡

It is easy to see,

Lemma 7.3 *There is a $1-1$ correspondence between A_n and $\mathcal{H}_n$.* ♡

Theorem 7.6 ($conv(A_n)$) *Given $n \geq 4$, and A_n as defined above, we have,*

1. *The polytope $conv(A_n)$ is full-dimensional, that is, $dim(conv(A_n)) = \alpha_n$*
2. *The barycentre of $conv(A_n)$ is given by,*

$$\bar{Y} = (\underbrace{1/p_3 1/p_3, \ldots,}_{2-times} \underbrace{1/p_{n-1} \ldots, 1/p_{n-1}}_{p_{n-1}-1 \; times}),$$

3. *$\bar{Y} \in int(conv(A_n))$.*

Proof Part 1 is Theorem 7.4.

Part 2 can be verified by noticing that the cardinality of $vert(conv(A_n))$, is $(n-1)!/2$, and in any X in P_n, the $(k-3)^{rd}$ component has p_{k-1} coordinates, and exactly one of the coordinates is a 1. In P_n for any component the 1 appears equally likely among the coordinates. And for any $Y \in A_n$ we have deleted the last coordinate in each component of the corresponding X.

The proof of part 3 of the theorem follows from the fact that $\bar{Y}$ does not lie on any facet defining hyperplane $CY = c_0$, for $(C, c_0) \in Q^{\alpha_n+1}$. Suppose, it lies on some facet defining hyperplane $CY = c_0$(, that is, $CY \leq c_0$ for all $Y \in conv(A_n)$). Then

$$C\bar{Y} - c_0 = [2/(n-1)!] \sum_{X \in P_n} (CY^X - c_0) = 0,$$

where Y^X is the element of A_n corresponding to a $X \in P_n$. Thus, for all $X \in P_n$, we have $CY^X = c_0$,

$$\implies dim(conv(A_n)) \leq \alpha_n - 1.$$

This contradicts the fact $dim(conv(A_n))$ is α_n. Therefore $\bar{Y} \in int(conv(A_n))$.

Hence the theorem. □

Theorem 7.7 (Facet—Complexity of $conv(A_n)$) *$conv(A_n)$ has facet complexity at most $\phi = 3\alpha_n^3 + 3\alpha_n^2(n-3)$. That is, $conv(A_n)$ is rationality guaranteed.* ♡

Proof Each vertex Y of $conv(A_n)$ is a $0-1$ vector of length α_n. So Y can be encoded with input size

$$\langle Y \rangle \leq \alpha_n + (n-3) = \nu.$$

(This follows from the fact that there are at most $(n - 3)$ $1's$ in any Y and $\langle 0 \rangle = 1$ & $\langle 1 \rangle = 1 + \lceil \log_2 2 \rceil = 2$.)

Therefore, $conv(A_n)$ has vertex complexity $\leq \nu$.

Using Lemma 7.1, we have, facet complexity of $conv(A_n)$,

$$\begin{aligned} &\leq 3\alpha_n^2 \nu \\ &= 3\alpha_n^2(\alpha_n + (n - 3)) \\ &= 3\alpha_n^3 + 3\alpha_n^2(n - 3). \end{aligned}$$

Hence, $conv(A_n)$ is rationality guaranteed.

Thus, we find $conv(A_n)$ satisfies all the requirements of Maurras's conditions (assumption stated in Sect. 7.3). Therefore, since we have a membership checking framework for $conv(P_n)$, which can be efficiently implemented in strongly polynomial time in n as shown in Chap. 6, we have an efficient membership oracle/subroutine for $conv(A_n)$. Given a $Y \notin conv(A_n)$, we can call the membership checking oracle a polynomial number of times to separate a $Y \in Q^{\alpha_n}$ from $conv(A_n)$.

We have identified a violated facet of $conv(A_n)$. Thus, we state this as the main theorem of the chapter.

Theorem 7.8 (Main Theorem) *Separation problem for* $conv(A_n)$ *has a polynomial time oracle.*   ♡

7.4 Conclusion

Given $C \in Q^{\alpha_n}$, consider the linear optimisation over $conv(A_n)$. Now applying GLS-Theorem, to the above result—showing there exists a polynomial separation oracle for $conv(A_n)$, we have established that linear optimisation over $conv(A_n)$ can be done in time polynomially bounded by α_n, ϕ, $\langle Y \rangle$, & $\langle C \rangle$. Or equivalently, linear optimisation over $conv(P_n)$ can be done in polynomial time, or *STSP* can be solved in polynomial time. Wow! That means the decision version of *STSP* which is NP-complete is in P, and hence from Cook's Theorem stated in Chap. 1 we have

$$NP = P.$$

7.5 Appendix: Excerpts from Maurras's Separation Construction for a Polytope

Let P be a well-defined polyhedron in the n-dimensional space $\mathbb{Q}^n$ of rational vectors, i.e., there is a bound k of the encoding length of any vertex of P. Additionally, suppose that P have a non-empty interior, and a point a in the interior of P is given. Notice that the encoding length of supporting hyperplanes of the facets of P and the encoding length of the vertices of P are polynomially equivalent. Let $\langle P \rangle$ denote an upper bound of the encoding length of the elements (vertices or facets) of P. Let $\mathrm{int}(P)$, $bd(P)$ be the interior and the boundary of P respectively.

Assume that P satisfies the assumption stated in Sect. 7.3. Given a point $x \notin P$ Maurras [121] presents an intuitive construction which, using a polynomial number of calls to the membership oracle finds a facet of P whose supporting hyperplane separates x from P.

We shall sketch this construction in the following section.

7.5.1 Construction and Its Validity

For a point $y \notin P$, the intersection of the segment $[a, y]$ with the boundary of P is denoted by $\underline{y}$. So $\underline{y}$ is the central projection with centre a of y on the boundary of P.

Let x be a point not in P of interest.

Denote by G the hyperplane perpendicular to the line (a, x) and passing through the point x. It is easy to see the equation of G is polynomially described in the encoding length. We will assume that our system of coordinates is composed of the direction (a, x) and a system of coordinates of G. Moreover, to avoid confusion between affine and linear independencies for points of $bd(P)$, we additionally assume that a is the origin of coordinates. The coordinate of the direction (a, x) is named the 1st , and the unit vector in the kth direction in G is denoted by v^k. The distance between x and P is of polynomial length, therefore we can choose a unit length for the coordinate system in G so that the points of G we subsequently construct will stay outside P.

Next, Maurass's construction involves two pairs of related sequences of points, $\left\{ \underline{x}^i, \underline{y}^i \right\}$ and $\left\{ x^i, y^i \right\}$, $i \in \{1, \ldots, n\}$. The points x^i, y^i belong to the plane G, while $\underline{x}^i$ $\underline{y}^i$ are their central projection with centre a on the boundary of P. Initially set $y^1 = x^1 = x$ and $\underline{y}^1 = \underline{x}^1 = \underline{x}$. Suppose that for $i \leq k$ we have constructed the points $\underline{x}^i$ and $\underline{y}^i$. Now, let us define the new points $\underline{x}^{k+1}$ and $\underline{y}^{k+1}$.

Definition of $\underline{x}^{k+1}$ and $\underline{y}^{k+1}$:

The two-dimensional plane spanned by the point a, the straight line (a, y^k) and the direction v^{k+1} intersects the polyhedron P along a polygon P^k.

Consider the intersection I^k of P^k with the triangle $T^k = \left(a, y^k, y^k + v^{k+1} \right)$, whose boundary is oriented by the vector v^{k+1}.

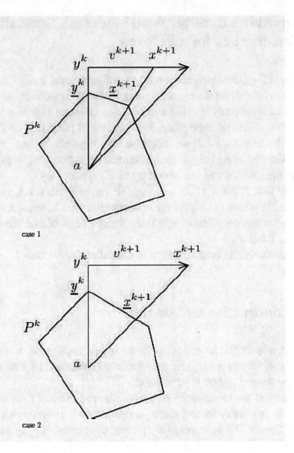

Fig. 7.1 Construction of x^{k+1}. *Source*: Maurras, J. F. (2002). From membership to separation, a simple construction. Combinatorica, 22(4):531–536

The point $\underline{x}^{k+1}$ is the first vertex of I^k, distinct from $\underline{y}^k$, in the direction of v^{k+1} (see Fig. 7.1). We distinguish two cases:

Case 1: the point $\underline{x}^{k+1}$ is a vertex of the polygon P^k not on the boundary of the triangle T^k (this is the case of points x_2 and $\underline{x}_2$ in Fig. 7.2);

Case 2: the point $\underline{x}^{k+1}$ is on the boundary of the triangle T^k (see the points x_3 and $\underline{x}_3$ in Fig. 7.2).

Set $\underline{y}^{k+1} := \frac{1}{2}\left(\underline{y}^k + \underline{x}^{k+1}\right)$

Maurras establishes the fact:

Lemma 7.4 *The points x^i, as well as $\underline{x}^i$, $i \in [n]$ are of polynomial encoding length.* ♡

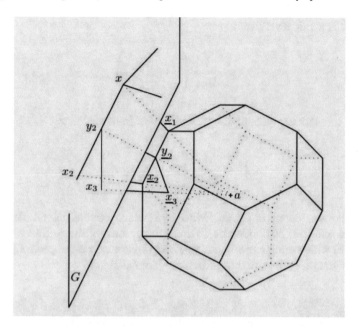

Fig. 7.2 General construction. *Source*: Maurras, J. F. (2002). From membership to separation, a simple construction. Combinatorica, 22(4):531–536

Proof The intersection of the hyperplane G and the polyhedral cone with origin a, basis F, and bounded below by the hyperplane parallel to G and passing through a, is a well-defined polyhedron P_F in G. The central projection on the hyperplane H of the basis of the space defined by the hyperplane G is a basis of H. However, if we change the origin of the basis in G, the corresponding basis in H will change. Therefore we will perform our construction simultaneously in G and H, the analysis being easier in G. In the polyhedron P_F, let us express x^{k+1} as a function of y^k. We have:

- either $x^{k+1} = y^k + \lambda v^{k+1}$
- or $x^{k+1} = y^k + v^{k+1}$, and this point has the same encoding length as y^k.

In the first case, the value of λ is chosen so that the point $y^k + \lambda v^{k+1}$ belongs to a facet of P_F. Let $\sum_{j=2}^{n} \alpha_j z_j = \beta$ be the equation of the supporting hyperplane of this facet written with integer coefficients. Then we have

$$\sum_{j=2}^{n} \alpha_j \left(y_j^k + \lambda v_j^{k+1} \right) = \beta$$

Since v^{k+1} is a unit vector, we have:

$$\lambda = \frac{1}{\alpha_{k+1}}\left(\beta - \sum_{j=2}^{n} y_j^k\right)$$

Only the $(k+1)$th component of x^{k+1} is different from that of y^k :

$$x_{k+1}^{k+1} = \frac{1}{\alpha_{k+1}}\left(\beta - \sum_{j=2}^{n} \alpha_j y_j^k\right)$$

For y^{k+1} the components are again divided by 2, which remains polynomial. The last component will thus be divided by 2^n times the product of the α_i (not of the same hyperplane) which remains polynomial. The $\underline{x}^i$ which are central projections of the x^i on the hyperplane H_F are thus also of polynomial length.

> **Theorem 7.9** (Maurras Separation Theorem) *The points $\underline{x}^1, \underline{x}^2, \ldots, \underline{x}^n$ span a supporting hyperplane H_F of a facet F containing the point $\underline{x}^1 = \underline{x}$, and thus separating x and the polytope P. These points and the hyperplane H_F can be obtained after a polynomial number of calls to the membership oracle.*
>
> ♡

Proof First notice that the section of P with the two-dimensional plane spanned by the points a and y^k, and the direction v^{k+1}, is the polygon P^k, therefore the points y^k and $\underline{x}^{k+1}$ are distinct. Notice also that the direction v^{k+1} in the hyperplane G is independent of the previous ones, thus the points $x^1, x^2, \ldots, x^n$ are affinely independent. Since the points $\underline{x}^1, \underline{x}^2, \ldots, \underline{x}^n$ belong to a facet F of P, they belong to the supporting hyperplane H_F of this facet. Moreover, $\underline{x}^i = \left[x^i, a\right] \cap H_F$, therefore these points are affinely independent as well.

Every face of P which contains a point inside a simplex $S \subseteq P$ also includes S itself. Thus a face which contains both y^k and $\underline{x}^{k+1}$ also contains all the $\underline{x}^i$ for $i \leq k + 1$. Thus the points $\underline{x}^i$ for $i = 1, 2, \ldots, n$ span the hyperplane H which supports the facet F containing $\underline{x}$, establishing the first assertion.

From Lemma 7.4, $x^i, \underline{x}^i$ are of polynomial encoding length, we can effectively construct, in polynomial time, the points $\underline{x}^i$ and y^i, by employing binary search.

Given the point x^i, on the segment $\left[a, x^i\right]$ by binary search using the membership oracle, we can polynomially approximate $\underline{x}^i$. Now, using a reduction by continued fractions on each component we can calculate the exact value of $\underline{x}^i$

Let z be on the hyperplane G outside of P and let z' be the middle of the segment $[y^k, z]$. We can test in polynomial time, if the segment $\left[y^k, \underline{z}\right]$ lies on the boundary of P. For this, we have just to check if the points $\underline{z}'$ and the intersection of the lines (a, z') and $\left(y^k, \underline{z}\right)$ are the same. By binary search on the segment $[y^k, y^k + v^{k+1}]$, and using the previous construction we can approximate x^{k+1}, and, using continuous fractions again, we can compute the point x^{k+1}.

Chapter 8
Epilogue

8.1 Introduction

Considering what has been theoretically established and concluded in Chap. 7, there could be nothing more fitting to be an epilogue. But it is a new beginning, and hence a glimpse of what we have seen, observed working on *MI*-formulation and pedigree optimisation for small to medium size problems, is reported in this chapter.[1] Problem instances considered in the computational experiments fall into one of the three categories, namely, (1) some randomly generated instances, (2) some from the library of sample instances for the *TSP* (and related problems), called TSPLIB [147], and (3) instances from [139] called diamond instances with different number of diamonds. In addition, with the view to solving larger instances of the *MI*-formulation, some new theoretical understanding of *MI*-relaxation is presented from [1] a hypergraph perspective, with the hope of solving and designing new algorithms for *MI*-relaxation, and [2] a Lagrangian relaxation perspective, noting the success of such approaches to difficult combinatorial problems in practice. Last but not least the chapter ends giving a reason for the definition of a pedigree appearing in Chap. 1 using triangles, and concludes the book.

8.2 Comparison of Different Formulations of STSP

Recall the polyhedral analysis mentioned in Sect. 3.4 in Chap. 3. Such polyhedral analyses have been used in the literature to compare several existing formulations of *TSP* based on the corresponding polyhedra involved [19, 78, 81, 132, 134, 136, 150, 165]. In addition to their comparative study using polytope analysis,

---

[1] These computational experiments were undertaken by Laleh Ardekani [90], and M. Gubb [86] years before the unexpected result of NP = P is established by the author in this book.

Öncan et al. [132] report some results on the empirical quality of the LP bounds obtained with some of the formulations. They tested these formulations on 10 *ATSP* instances with the number of cities ranging from 30 to 70, obtained from TSPLIB [147]. They use the relative deviations from the optimal tour lengths, computed as $100 \times (z_{IP}^* - z_{LP}^*)/z_{IP}^*$, where z_{IP}^* is the length of an optimal tour and z_{LP}^* is the lower bound obtained by solving the LP relaxation of the models. In their study, the formulation of Sherali, Sarin, and Tsai (*SST2*) [156] and the one given by Claus (*CLAUS*) [46] provide the best bounds. These comparisons are primarily for *ATSP* formulations.

Gubb [86] considers the polynomial formulations discussed in [132] and three new symmetric formulations and presents computational comparisons on these formulations, including *MI*-formulation. Randomly generated Euclidean *TSP* instances with $n = 100$ were tested along with 10 instances, with the number of cities ranging from 30 to 70 from TSPLIB [147]. These results were compared with the implications of the polyhedral analysis, and the results show that polyhedral analysis by itself does not necessarily show which formulations are faster in practice. The polyhedral analysis gives one aspect, the strength of the LP relaxation, without any insight into the time taken to solve the LP relaxation. Practical comparisons need both, and to link the relationship between them to fully solve for an optimal integral solution.

Gubb concludes based on his computational results that the best polynomial *STSP* formulation is Multistage Insertion formulation compared to the formulations studied in [132]. In general, *MI*-formulation fully solved the instances and is the fastest; as it is not only as strong as *DFJ* formulation but also has a small number of variables compared to other polynomial formulations that are equivalent to *DFJ*-formulation.

Papadimitriou and Steiglitz [139] designed a series of *STSP* instances to show how some local search algorithms, in particular the k–interchange heuristics, get stuck in local optima and fail to find the optimum. All of these instances have a single global optimum but exponentially many second-best local optima that are all appreciably inferior to the global optimum. The instances are made up of some structures called diamonds and are indexed by the number of diamonds.

Ardekani [90] compares the performance of LP relaxations of various *TSP* formulations with that of *MI*-formulation. Some *STSP* and *ATSP* instances from TSPLIB and some diamond instances by Papadimitriou and Steiglitz [139] are used for this purpose. *MI*-relaxation outperforms other formulations either in terms of solution time and the number of iterations, or the quality of the solutions, on *STSP* instances from TSPLIB. The solution times and the number of Cplex iterations for the *MI*-formulation are significantly lower than those of other formulations with the same LP relaxation value. Despite their lower solution times, the gap with the optimal solution given by the formulation of Fox, Gavish, and Graves [74], and that of Miller, Tucker, and Zemlin [123] are considerably higher than that of *MI*-formulation. The *MI*-formulation performs better than formulations given by Carr [40], Claus [46], and Flood [71] in terms of solution time and number of iterations on *TSP* instances from TSPLIB. Ardekani notes that for the *ATSP* instances reported by Gouveia and Pires [80], the gap with the optimal solution for the multicommodity flow (MCF) formulation dominates that of the *MI*-formulation; however, solution times required by this formulation are notably greater than those of the *MI*-formulation.

Ardekani observes that LP relaxation of *MI*-formulation found the integer solution to all the *STSP* diamond instances. *MI*-formulation also outperforms other formulations in terms of LP relaxation value or solution time [8]. Some of the tables from Ardekani [90] are reproduced in Sect. 8.8.2 as Appendix 2.

Roberto and Toth [150] show that the branch and cut code *FLT*, by Fischetti et al. [69] and *Concorde* code by Applegate et al. [5, 50], were the only two exact methods based on the branch and cut polyhedral approach, that can solve, within the imposed time limit, all the 35 instances from TSPLIB and some real world instances to optimality, and are better performing than the other five branch and bound methods considered by them. The largest problem size solved by them is 443. Also, they conclude that *Concorde* is the only code available to solve very large-size instances. *Concorde* and *FLT* work in the space of variables used in *DFJ* formulation. And such a cutting plane approach needs to ensure the algorithm can check whether any subtour elimination constraint is violated and it must be able to solve the LP relaxation efficiently as well. The facet (and adjacency) structure of the tour polytope contained in the subtour elimination polytope plays an important role in the computational difficulty faced by these cutting plane methods. Hence, a new beginning is possible if we choose, instead of working with tours, to work with pedigrees [15].

The main aim of this chapter is to provide new algorithmic approaches that might appreciably increase the size of problems solved using *MI*-relaxation. This is the first step towards taking a different polyhedral approach based on the pedigree polytope to solve the *STSP* problem. Thus, we are not at this stage competing with *Concorde*, but suggest methods based on new results on *MI*-formulation to solve larger instances of the *MI*-relaxation problem. That is, the first step in using *MI*-polytope that contains pedigree polytope, instead of, the standard polytope that contains Q_n, to solve *STSP*.

Though *MI*-relaxation has a strongly polynomial theoretical algorithm to solve, since it is a combinatorial LP, we explore further properties of the *MI*-relaxation problem, to enable solving larger instances of *STSP* using the pedigree approach.

The next three subsections give a preliminary account of the concepts and tools used in the hypergraph flow problem, Lagrangian relaxation approach and Leontief substitution systems, to derive new results on *MI*-relaxation.

8.2.1 Hypergraph and Flows

Here we state some of the definitions and concepts and related works by Italian researchers, Riccardo Cambini, Giorgio Gallo, Maria Grazia Scutellà among others [38, 39].

Definition 8.1 A directed hypergraph is a pair $\mathcal{H} = (\mathcal{V}, \mathcal{E})$, where $\mathcal{V} = \{v_1, v_2, \ldots, v_n\}$ is the set of vertices, and $\mathcal{E} = \{e_1, e_2, \ldots, e_n\}$ is the set of *hyperarcs*. A hyperarc e is a pair (T_e, h_e), where $T_e \subset \mathcal{V}$ is the tail of e and $h_e \in \mathcal{V} \setminus T_e$ is its head. A hyperarc that is headless, $(T_e, \emptyset)$, is called a *sink* and a tailless hyperarc, $(\emptyset, h_e)$, is called a *source*.   ♣

Given a hypergraph $\mathcal{H} = (\mathcal{V}, \mathcal{E})$, a positive real multiplier $\mu_v(e)$ associated with each $v \in T_e$, a real demand vector b associated with $\mathcal{V}$, and a nonnegative capacity vector w, a flow in $\mathcal{H}$ is a function $f : \mathcal{E} \to R$ which satisfies:

$$\sum_{e \ni v = h_e} f(e) - \sum_{v \in T_e} \mu_v(e) f(e) = b(v), \ \forall v \in \mathcal{V} \qquad \text{(conservation)}, \qquad (8.1)$$

$$0 \leq f(e), \ \forall e \in \mathcal{E} \qquad \text{(feasibility)}, \qquad (8.2)$$

$$f(e) \leq w(e), \ \forall e \in \mathcal{E} \qquad \text{(and capacity)}. \qquad (8.3)$$

Problem 8.1 (*Minimum cost hypergraph flow problem*) Let $c(e)$ be the cost associated with the hyperarc e, $\forall e \in \mathcal{E}$. Find f^* such that $\sum_{e \in \mathcal{E}} c(e) f^*(e)$ is a minimum over all f satisfying the flow constraints (8.1), (8.2), and (8.3).

Notice that this problem generalises not only the minimum cost flow problem in ordinary networks but also the minimum cost generalised flow problem [133]. A directed path P_{st} from s to t in $\mathcal{H}$ is a sequence $P_{st} = (v_1 = s, e_1, v_2, e_2, \ldots, e_q, v_{q+1} = t)$, where $s \in T_{e_1}$, $h_{e_q} = t$, and $v_i \in T_{e_i} \cap h_{e_{i-1}}$ for $i = 2, \ldots, q$.

If $s = t$, then P_{st}, is a *directed cycle*. When no directed cycle exists, then $\mathcal{H}$ is called a *cycle-free* hypergraph. A directed hyperpath, Π_{St} from the source set S to the sink t in $\mathcal{H}$ is a minimal cycle-free sub-hypergraph containing both the nodes in S and node t, and such that each node, except for the nodes in S has exactly one entering hyperarc. A hyperarc e' is said to be a *permutation* of a hyperarc e if $T_e \cup \{h_e\} = T_{e'} \cup \{h_{e'}\}$. A hypergraph $\mathcal{H}'$ is a permutation of a hypergraph $\mathcal{H}$ if its hyperarcs are permutations of the hyperarcs of $\mathcal{H}$. A *directed hypertree* with root set R, and a set of hyperarcs, E_T, called tree arcs, is a hypergraph $\mathcal{T}_R = (R \cup N, E_T)$ such that:

1. $\mathcal{T}_R$ has no isolated vertices, and does not contain any directed cycle;
2. $R \cap N = \emptyset$;
3. Each node $\nu \in N$ has exactly one entering hyperarc; and
4. No hyperarc has a vertex of R as its head.

We say $\mathcal{T}_R$ is a directed hypertree rooted at R and N is called the set of *non-root* vertices. Any non-root vertex not contained in the tail of any tree arc is called a *leaf*. Any permutation of a directed hypertree rooted at R yields an undirected hypertree rooted at R. It can be shown that $\mathcal{T}_R$ is a directed hypertree rooted at R if and only if

- $\mathcal{T}_R$ has no isolated vertices,
- $R \cap N = \emptyset$ and $|N| = |E_T| = q$,
- an ordering $(v_1, v_2, \ldots, v_q)$ and $(e_1, e_2, \ldots, e_q)$ exists for the elements of N such that: $h_{e_j} = v_j$, and
- $R \cup \{v_1, v_2, \ldots, v_{j-1}\} \supseteq T_{e_j}, \forall e_j \in E_T$.

8.2.2 Lagrangian Relaxation and Variants

This section briefly collects relevant results and concepts from the Lagrangian relaxation approach to finding lower bounds for difficult combinatorial problems. The Lagrangian relaxation approach has found applications in several different areas of practical problems of scheduling (mixed integer and integer programming formulations of the problems are relaxed using this approach). Guignard [87] gives many examples of this nature explaining various new developments and practical implementation tips. The crucial research on Lagrangian relaxation for combinatorial optimisation problems were the seminal papers for *TSP* by Held and Karp [92, 93]. Firstly, they showed the Lagrangian relaxation based on 1-tree for *TSP* gives the same bound as the LP relaxation of the standard formulation of *TSP*.

Subsequently Held and Karp introduced subgradient optimisation to solve Lagrangian duals. Computational experiments with subgradient optimisation are known to be successful. Important in Held and Karp's use of Lagrangian relaxation is the creative identification of the subproblem of solving 1-tree minimisation, which breaks down to solving a minimum spanning tree problem, that can be efficiently solved. In addition, the lower bound obtained (by solving Lagrangian dual) using 1-tree relaxation, is a tight lower bound for *TSP*, which goes by the name *HK-bound* [161]. For solving large-scale convex optimisation problems and for providing lower bounds on the optimal value of discrete optimisation problems, researchers have effectively used Lagrangian relaxation and duality results. A key approach in providing efficient computational means to obtain near-optimal dual solutions and bounds on the optimal value of the original problem, since the work of Held and Karp [93], has been *subgradient optimisation* or methods based on subgradients. Geoffrion's paper [77] is an early work on Lagrangian relaxation and its use in $0-1$ programming problems. (Several other references can be found in [87]).

But the computational complexity of subgradient optimisation is not known to be polynomial, however, Bertsimas and Orlin in [31] showed the existence of fast polynomial algorithms for solving Lagrangian dual of LP problems, in some structured LP instances guaranteeing solution to the primal problem as well. Theoretically, their method is significantly faster than interior point methods and ellipsoid-like methods directly applied to the problem.

Next, we give the definitions and results for use in later parts of the paper. Let P be the problem $\min\{cx \mid Ax \leq b, Dx \leq d, x \geq 0, x \in \mathcal{X}\}$. Let $F(P)$ denote the set of feasible solutions to a given problem P. Let $F^*(P)$ be the set of optimal solutions to P. Let $z(P)$ be the optimal value of the problem P. Usually $\mathcal{X}$ imposes some integrality restrictions on the solutions of P.

Definition 8.2 A Lagrangian relaxation of (P) is either

$$\min_{x}\{cx + \lambda(Ax - b) \mid Dx \le d, x \in \mathcal{X}\}$$

or

$$\min_{x}\{cx + \lambda(Dx - d) \mid Ax \le b, x \in \mathcal{X}\}$$

depending on which among the sets of constraints is relaxed using nonnegative λ, called Lagrangian multipliers.   ♣

To fix ideas, we generally relax those constraints that make the resulting Lagrangian problem easy to solve, we call the remaining constraints, *kept constraints*. Sometimes, both sets, when relaxed may result in relatively easier Lagrangian problems. However, both relaxations provide lower bounds on the optimal value of P, $z(P)$. Let us assume, $Ax \le b$ are relaxed unless otherwise specified.

The problem of finding the supremum of such lower bounds leads us to the Lagrangian Dual problem. Let us call LR_λ, the Lagrangian relaxation problem under consideration.

Definition 8.3 The LD problem

$$z(LD) = \max_{\lambda \ge 0} z(LR_\lambda)$$

is called the *Lagrangian Dual* of P, relative to the set of constraints relaxed, $Ax \le b$.   ♣

Let $x(\lambda)$ denote an optimal solution of LR_λ for some $\lambda \ge 0$. If $x(\lambda)$ is feasible for P, then $cx(\lambda) + \lambda(Ax(\lambda) - b) \le z(P) \le cx(\lambda)$. If in addition, $\lambda(Ax(\lambda) - b) = 0$ then $x(\lambda)$ is an optimal solution to P, as well, i.e., $z(P) = cx(\lambda)$. However, if instead of $Ax \le b$, we had $Ax = b$, set of equality restrictions, then λ is not required to be nonnegative in Lagrangian problem. In that case, if $x(\lambda)$ is optimal for Lagrangian problem and is feasible for P then it is optimal for P as well, $z(P) = cx(\lambda)$.

The Lagrangian relaxation bound is never worse than the LP bound. However, if $conv\{x \in \mathcal{X} \mid Dx \le d\} = \{x \mid Dx \le d\}$, then $z(LP) = z(LD) \le z(P)$. This is an important result as we don't hope to improve the bound if the relaxed problem has integral extreme points. Thus, the reason we relax them emanates from the fact that solving the LP is hard either due to the number of constraints (exponentially many) or the basis size (too large for the solver). In such cases, solving the LD becomes an alternative approach to calculating the LP bound.

In general, Lagrangian solutions, which are feasible integral for the kept constraints may violate one or more of the relaxed constraints. In such a situation, Lagrangian solutions could be used as input for some heuristics that achieve feasibility for the problem P. Barahona and Anbil [26] give an averaging method to recover feasible solutions to the primal problem from the solutions to Lagrangian problems.

Suppose $\mathcal{C} = conv\{x \in \mathcal{X} \mid Dx \le d\}$ is a polytope, then $\mathcal{C}$ has finitely many extreme points, $x^s, s \in S$, such that

$$z(\lambda) = \min\{cx + \lambda(Ax - b) \mid Dx \le d, x \in \mathcal{X}\} = \min_{s \in S}\{cx^s + \lambda(Ax^s - b)\}.$$

$z(\lambda)$ is a piecewise linear concave function, so Lagrangian dual seeks a maximum of a concave function, $\max z(\lambda)$, that is not differentiable everywhere. At any breakpoint λ^0 of $z(\lambda)$, we have subgradients $y \ni$

$$z(\lambda) - z(\lambda^0) \le \langle y, (\lambda - \lambda^0) \rangle,$$

where $\langle y, (\lambda - \lambda^0) \rangle = \sum_i y_i(\lambda_i - \lambda^0)$.

The set of all subgradients of a concave function $z(\lambda)$ at a point (λ^0) is called the subdifferential of the function z at the point λ^0 and is denoted by $\partial z(\lambda^0)$. From convex analysis we know the following facts, which we state without proof:

Lemma 8.1 *The subdifferential $\partial z(\lambda^0)$ is a nonempty, closed, convex and bounded set. And if $\partial z(\lambda^0)$ is a singleton set then that is the gradient of $z(\lambda)$ at λ^0.* ♡

Lemma 8.2 *If $z(\lambda)$ is not differentiable at λ^k, then $y^k = (Ax^k - b)^T$ is a subgradient of $z(\lambda)$ at λ^k. y^k is orthogonal to $\{\lambda \mid cx^k + \lambda(Ax^k - b) = \eta^k\}$ where $\eta^k = z(\lambda^k)$, the optimal value for Lagrangian problem corresponding to λ^k.* ♡

To solve LD, Lagrangian dual, Held and Karp used subgradient optimisation, in which iteratively LR_λ problems are solved for a sequence of λ_k, choosing the next λ_{k+1} moving a step along a subgradient of $z(\lambda)$ at λ_k. The step size is an important issue for practical convergence of the subgradient optimisation method, though theoretically Poljak [140, 141] has shown that given $t_i \in R, i \in N \ni \sum_{i=1}^{\infty} t_i = \infty$ and $\lim_{i \to \infty} t_i \to 0$; and $\lambda_{i+1} = t_i \lambda_i$, we have $\lim_{\lambda_i \to \infty} z(LR_{\lambda_i}) = z(LD) = \max_{\lambda \ge 0} z(LR_\lambda)$.

The choice

$$\lambda_{k+1} = \lambda_k + \frac{s^k \, \epsilon_k (\eta^* - \eta^k)}{||s^k||^2}$$

where s^k is the subgradient of $z(\lambda)$ at λ^k; $\epsilon_k \in (0,2)$, and η^* is an estimate of the optimal value of Lagrangian dual LD, was suggested by Held and Karp [93].

Other researchers (like Reinelt [146], Valenzuela and Jones [161]), have given alternative step size choices. Subgradient methods use different step-size rules to update Lagrangian multipliers to ensure convergence. For solving nondifferentiable problems Poljak [141] initiated such studies on their convergence properties; under various step-size rules, convergence rates have been established by subsequent researchers. Nedić and Ozdaglar [131] report on recent great successes of subgradient methods in networking applications. They study methods for generating approximate primal solutions as a byproduct of subgradient methods applied to Lagrangian duals; they also provide theoretical bounds on the approximation. Recently Zamani and Lau [167] and Lorena and Narciso [114] have used surrogate information and learning capability, respectively, in Lagrangian relaxation to outperform Held and Karp's subgradient approach for solving Lagrangian duals. Using step size depending on the previous iterations, with the capacity of expansion or contraction of the step size, Zamani and Lau solve Euclidean *TSP* instances from *TSPLIB* [147] and claim that the procedure was very effective.

To solve Lagrangian dual, in addition to subgradient methods and their variations, there are other approaches like the analytic centre cutting plane method, volume algorithm, and Bundle method [87, 111]. As cited earlier, Barahona and Anbil [26] give an averaging method called volume algorithm to recover feasible solutions to the primal problem from the solutions to Lagrangian problems.

8.2.3   *Leontief Substitution Flow Problems*

First, some definitions from Veinott [162], the classic work on Leontief matrices. Consider $A \in R^{m \times n}$, $b \in R^m$ and $x \in R^n$.

Definition 8.4 (*trivial rows*) The ith row (column) of a matrix A is called *trivial* if for every column vector $x \geq 0$ for which $Ax \geq 0$, the ith component of Ax (x) is zero; otherwise the ith row (column) is called *nontrivial*. ♣

Definition 8.5 (*pre-Leontief, Leontief matrices*) A matrix A is called *pre-Leontief* if each column has at most one positive element. A matrix is called Leontief if each column has exactly one positive element and its rows are nontrivial. ♣

Definition 8.6 (*pre-Leontief, Leontief Substitution Systems*) The system $Ax = b, x \geq 0$ with $b \geq 0$ is called pre-Leontief or Leontief substitution system according as the given matrix is pre-Leontief or Leontief. ♣

Remark

1. So the incidence matrix of a hypergraph is pre-Leontief.
2. The hypergraph flow problem is a pre-Leontief substitution system in case the demand vector b is nonnegative. If in addition, the rows are nontrivial, we have a corresponding Leontief substitution system.
3. Conversely it is easy to see that given a pre-Leontief system, we can define a corresponding hypergraph flow problem as defined in Sect. 8.2.1. (Lemma 1.1, [99]).
4. The term *degenerate* hyperarc is used in [99] if $x(e) = 0$ corresponding to a hyperarc e in every basic feasible solution to the problem. (Trivial column as in Definition 8.4.)

Definition 8.7 (*Leontief substitution flow problem* [99]) Given $C \in R^{|\mathcal{E}|}$, and a Leontief matrix A, we call the minimum cost hypergraph flow problem given by minimise $\{Cf \mid Af = b, f \geq 0\}$ a Leontief substitution flow problem, in case $b \geq 0$. ♣

Definition 8.8 (*Jeroslow* [99]) Consider a hypergraph $\mathcal{H}$. Let $v_1, e_1, v_2, e_2, \ldots, e_k, v_{k+1}$ be a directed cycle where $v_{k+1} = v_1$ and $e_i = (T_i, v_{i+1})$, and $v_{i-1} \in T_i, i = 1, \ldots, k$. The gain of this directed cycle is defined by

$$1/\Pi_{i=1}^{k}\mu_{v_i}(e_i).$$

We call a Leontief flow problem defined on a hypergraph $\mathcal{H}$, gain free if the gain of every directed cycle in $\mathcal{H}$ is ≤ 1. ♣

Lemma 8.3 *If the incidence matrix of a hypergraph $\mathcal{H}$ is integral then a Leontief flow problem defined on the hypergraph $\mathcal{H}$, is gain-free.*   ♡

Proof The proof follows from the definition of gain.

The integrality of the basic feasible solutions to Leontief substitution flow problems is known from Veinott's classical result:

Theorem 8.1 (Veinott [162]) *Let $Ax = b$, $x \geq 0$, be a given system. If A is an integral Leontief matrix, the following are equivalent:*

1. *B^{-1} is integral for every $B \in \mathcal{B}^*$,*
2. *$det(B) = 1$ for every $B \in \mathcal{B}^*$,*
3. *The extreme points of $F(b)$ are integral for every nonnegative integral b,*

where (a) $F(b) = \{x \mid Ax = b, x \geq 0\}$, denotes the solution set of the given system, (b) for a pre-Leontief matrix A, let $\mathcal{B}$ be the set of all square full submatrices of A having pre-Leontief transpose and (c) $\mathcal{B}^*$ is the set of all Leontief matrices contained in $\mathcal{B}$ (for a proof see [162], p.17).   ♡

From the definition of unimodular matrices, we know that a unimodular matrix B is a square integer matrix with determinant a $+1$ or -1. And we know that a matrix A is totally unimodular if every square non-singular submatrix is unimodular, i.e., every subdeterminant of A is either $+1$, -1, or 0. So if A is a totally unimodular matrix then $a_{ij} = 0$, or ± 1. So pre-Leontief matrices do not satisfy this requirement in general. However, we can understand the implication of Veinott's result better by using the definition of totally dual integrality (*TDI*) of linear programming problems.

Definition 8.9 (*Totally Dual Integral* [66]) Given an integral A and rational b the LP $min\{cx \mid Ax = b, x \geq 0\}$ is totally dual integral (*TDI*) if the dual problem has an integral optimal solution for every integer vector c for which it has an optimal solution.   ♣

Jeroslow et al. [99] have shown that every Leontief substitution flow problem with an integral A and rational b is *TDI*. And using this they prove that if a Leontief substitution flow problem with integral A, b has an optimal solution then it has an integral optimal solution. Moreover, they indicate how this result can be derived from Veinott's Theorem 8.1.

8.3   MI-Relaxation and Leontief Substitution Flow Problem

A special case of the minimum cost hypergraph flow problem discussed in Sect. 8.1 is
the minimum cost gain-free Leontief substitution flow problem. In this section, I show
that *MI*-formulation contains a subproblem, that is, a gain-free Leontief substitution
flow problem.

Consider a hypergraph $\mathcal{H} = (\mathcal{V}, \mathcal{E})$. We can write the vertex-hyperarc incidence
matrix $M = [m_{v,e}]$ corresponding to $\mathcal{H}$ as follows:

$$m_{v,e=(T_e,h_e)} = \begin{cases} 1 & \text{if } v = h_e, \\ -\mu_v(e) & \text{if } v \in T_e, \\ 0 & \text{otherwise.} \end{cases}$$

We can write the minimum cost hypergraph flow Problem 8.1 equivalently in
matrix form as:

$$\min \sum_{e \in \mathcal{E}} c(e) f(e) \tag{8.4}$$

$$M f = b, \tag{8.5}$$

$$f \geq 0, \text{ and} \tag{8.6}$$

$$f \leq w. \tag{8.7}$$

Notice that in M every column corresponding to a hyperarc has at most one
positive element, namely, $+1$. Such matrices are studied in economics, inventory
management, and other fields.

Recall the matrix notation of *MI*-formulation given using Definition 4.1 (see
Figs. 8.1 and 8.2 given for $n = 6$).

As seen earlier, *MI*- formulation can be recast in matrix notation as:

$$\text{Find } X^* \ni CX^* = \min \left\{ CX \mid X \ni E_{[n]}X = \mathbf{1}_{n-3}, A_{[n]}X \leq \begin{bmatrix} \mathbf{1}_3 \\ 0 \end{bmatrix}, X \in \{0, 1\}^{T_n} \right\}.$$

Observe that the elements of the constraint matrix of *MI*-formulation are $0, \pm 1$;
the right-hand side is a nonnegative integer vector and there are two $+1$'s in each

$$E_{[6]} = \begin{pmatrix} 1 & 1 & 1 & 0 & 0 & 0 & 0 & 0 & 0 & 0 & 0 & 0 & 0 & 0 & 0 & 0 & 0 & 0 \\ 0 & 0 & 0 & 1 & 1 & 1 & 1 & 1 & 1 & 0 & 0 & 0 & 0 & 0 & 0 & 0 & 0 & 0 \\ 0 & 0 & 0 & 0 & 0 & 0 & 0 & 0 & 0 & 1 & 1 & 1 & 1 & 1 & 1 & 1 & 1 & 1 \end{pmatrix},$$

Fig. 8.1   Matrix $E_{[n]}$ for $n = 6$

$$
A_{[6]} = \left(
\begin{array}{rrrrrrrrrrrrrrrrrrr}
1 & 0 & 0 & 1 & 0 & 0 & 0 & 0 & 0 & 1 & 0 & 0 & 0 & 0 & 0 & 0 & 0 & 0 & 0 \\
0 & 1 & 0 & 0 & 1 & 0 & 0 & 0 & 0 & 0 & 1 & 0 & 0 & 0 & 0 & 0 & 0 & 0 & 0 \\
0 & 0 & 1 & 0 & 0 & 1 & 0 & 0 & 0 & 0 & 0 & 1 & 0 & 0 & 0 & 0 & 0 & 0 & 0 \\
-1 & -1 & 0 & 0 & 0 & 0 & 1 & 0 & 0 & 0 & 0 & 0 & 1 & 0 & 0 & 0 & 0 & 0 & 0 \\
-1 & 0 & -1 & 0 & 0 & 0 & 0 & 1 & 0 & 0 & 0 & 0 & 0 & 1 & 0 & 0 & 0 & 0 & 0 \\
0 & -1 & -1 & 0 & 0 & 0 & 0 & 0 & 1 & 0 & 0 & 0 & 0 & 0 & 1 & 0 & 0 & 0 & 0 \\
0 & 0 & 0 & -1 & -1 & 0 & -1 & 0 & 0 & 0 & 0 & 0 & 0 & 0 & 1 & 0 & 0 & 0 & 0 \\
0 & 0 & 0 & -1 & 0 & -1 & 0 & -1 & 0 & 0 & 0 & 0 & 0 & 0 & 0 & 1 & 0 & 0 & 0 \\
0 & 0 & 0 & 0 & -1 & -1 & 0 & 0 & -1 & 0 & 0 & 0 & 0 & 0 & 0 & 1 & 0 & 0 & 0 \\
0 & 0 & 0 & 0 & 0 & 0 & -1 & -1 & -1 & 0 & 0 & 0 & 0 & 0 & 0 & 0 & 1 & 0 & 0 \\
0 & 0 & 0 & 0 & 0 & 0 & 0 & 0 & 0 & -1 & -1 & 0 & -1 & 0 & 0 & -1 & 0 & 0 & 0 \\
0 & 0 & 0 & 0 & 0 & 0 & 0 & 0 & 0 & -1 & 0 & -1 & 0 & -1 & 0 & 0 & -1 & 0 & 0 \\
0 & 0 & 0 & 0 & 0 & 0 & 0 & 0 & 0 & -1 & -1 & 0 & 0 & -1 & 0 & 0 & 0 & -1 & 0 \\
0 & 0 & 0 & 0 & 0 & 0 & 0 & 0 & 0 & 0 & 0 & -1 & -1 & -1 & 0 & 0 & 0 & 0 & -1 \\
0 & 0 & 0 & 0 & 0 & 0 & 0 & 0 & 0 & 0 & 0 & 0 & 0 & 0 & -1 & -1 & -1 & -1 & \\
\end{array}
\right).
$$

Fig. 8.2 Matrix $A_{[n]}$ for $n = 6$

column corresponding to any x_{ijk}, once in $E_{[n]}$ and another time in $A_{[n]}$. And so, unfortunately, this problem does not correspond to a gain-free Leontief substitution flow problem.

Definition 8.10 We say X, a real vector, is a *pre-solution* to *MI*-relaxation if X satisfies the inequalities of the *MI*-relaxation problem other than nonnegativity restrictions. In addition, if X is nonnegative we say X is a *feasible pre-solution*. Equivalently, X is a pre-solution to *MI*-relaxation problem means

$$
A_{[n]}X + I_{P_n}u = \begin{bmatrix} \mathbf{1}_3 \\ 0 \end{bmatrix}.
$$

So a feasible presolution X, if in addition satisfies the equality constraints, $E_{[n]}X = \mathbf{1}_{n-3}$, we have a feasible solution to *MI*-relaxation. ♣

Define

$$
\mathcal{X}(n) = \{X \mid E_{[n]}X = \mathbf{1}_{n-3}\} \tag{8.8}
$$

$$
\mathcal{U}(n) = \{u \in R^{P_n}, u \geq 0 \mid \exists\, X \in \mathcal{X}(n) \ni A_{[n]}X + I_{P_n}u = \begin{bmatrix} \mathbf{1}_3 \\ 0 \end{bmatrix}, X \geq 0\} \tag{8.9}
$$

Remark If X is a pre-solution to *MI*-relaxation, then pre-multiplying both sides of $A_{[n]}X + I_{p_n}u = \begin{bmatrix} \mathbf{1}_3 \\ 0 \end{bmatrix}$ by c and noting that $cA_{[n]}$ is indeed $-C$, we get $cu = c\begin{bmatrix} \mathbf{1}_3 \\ 0 \end{bmatrix} + CX$. So, it is seen that the objective function of *MI*-formulation is equivalent to cu but for the constant $c_{12} + c_{13} + c_{23}$. Thus, we can equivalently write *MI*-relaxation problem as Problem 8.2.

Problem 8.2

$$\text{minimise } cu \tag{8.10}$$

$$\text{subject to } A_{[n]}X + I_{p_n}u = \begin{bmatrix} \mathbf{1}_3 \\ 0 \end{bmatrix} \tag{8.11}$$

$$X \geq 0 \tag{8.12}$$

$$u \geq 0 \tag{8.13}$$

$$X \in \mathcal{X}(n). \tag{8.14}$$

Now notice that the system given by Eqs. 8.11, 8.12 and 8.13 corresponds to a gain-free Leontief substitution system, as (a) the system matrix, $A_{[n]}$, I_{p_n} is a pre-Leontief matrix, (b) all the rows of this system are nontrivial, (c) elements of the system matrix are ± 1 or 0, (hence gain free) and (d) the right-hand side is a nonnegative integer vector. Therefore, a feasible presolution X to MI-relaxation with the corresponding u provides us with a solution (X, u) to this system. And the basic feasible solutions of this system are integral, by an application of Veinott's Theorem 8.1 and using the directed hypertree corresponding to the basis. Thus we have Theorem 8.2.

Theorem 8.2 *Consider Problem 8.2 excluding the restriction* (8.14)

$$\text{minimize } cu, \text{ subject to } A_{[n]}X + I_{p_n}u = \begin{bmatrix} \mathbf{1}_3 \\ 0 \end{bmatrix}, X, u \geq 0.$$

We have a gain-free Leontief substitution flow problem, and every feasible basis is integral. ♡

I shall use this fact subsequently to devise algorithms based on the Lagrangian relaxation of the equality constraints (8.14). In Sect. 8.2.2 we reviewed the literature on Lagrangian relaxation. I apply some of the approaches and tips to solve Problem 8.2.

I next propose different algorithms based on Lagrangian relaxation as applied to the MI-relaxation problem.

8.3.1 Lagrangian Relaxation of Multiple Choice Constraints

Consider Problem 8.2. We could rewrite it using Lagrangian multipliers corresponding to the multiple choice constraints given by Eq. (8.14) as:

Problem 8.3 (*Lagrangian Relaxation problem*) LR_λ

$$\text{minimise } cu + \sum_{k=4}^{n} \lambda_k \left(\sum_{1 \le i < j < k} x_{ijk} - 1 \right) \qquad (8.15)$$

$$\text{subject to } A_{[n]}X + I_{p_n}u = \begin{bmatrix} 1_3 \\ 0 \end{bmatrix} \qquad (8.16)$$

$$X \ge 0 \qquad (8.17)$$

$$u \ge 0 \qquad (8.18)$$

$$\lambda \text{ real.} \qquad (8.19)$$

As observed earlier in Theorem 8.2, Problem-LR_λ is a gain-free Leontief flow problem and so we can apply the strongly polynomial value iteration algorithm from [99]. Thus Lagrangian subproblem for a given λ is not only easy to solve but also provides an integer optimal solution. Now to solve Lagrangian duals, LD, we could resort to subgradient optimisation or other approaches.

8.3.2 Value Iteration Algorithm for Problem 8.3

Consider the dual of the Problem LR_λ for a given λ. We introduce the variable π_{ij} for each row of the Eq. 8.16, $\forall\, 1 \le i < j \le n$. For each hyperarc $(\{(i, k), (j, k)\}, (i, j))$ corresponding to the variable x_{ijk} we have a dual constraint; for each hyperarc $(\emptyset, (i, j))$ corresponding to the slack variable u_{ij} we have an upper bound constraint for the dual variable, π_{ij} as follows:

$$\pi_{ij} \le \lambda_k + \pi_{ik} + \pi_{jk}, \; \forall\, 1 \le i < j < k; \; k = 4, \ldots, n, \qquad (8.20)$$

$$\pi_{ij} \le c_{ij}, \; \forall 1 \le i < j \le n. \qquad (8.21)$$

In [99] we have a strongly polynomial successive approximation scheme. This is based on the Bellman–Ford method (see, for instance, Chap. 7 of [108]), which finds either a negative cost cycle or $s - t$ shortest paths for a given $s \in G$, a directed graph with edge costs. We specialise in the value iteration algorithm for this specific gain-free Leontief flow problem at hand and get Algorithm 3. Let M be a sufficiently large positive value, which is used to initialise the solution.

Lemma 3.2 of [99] asserts that given a Leontief flow problem for a directed hypergraph $\mathcal{H} = (\mathcal{V}, \mathcal{E})$, if there is a basic feasible flow $\bar{f}$ having acyclic support with a hyperarc directed into every nontrivial vertex, and a complementary dual solution feasible for the dual, then for all nontrivial vertex in $\mathcal{V}$, their value iteration algorithm finds this dual solution after at most $|\mathcal{V}|$ iterations. So we have a strongly polynomial method for solving the problem LR_λ. We can recover the primal solution for the problem LR_λ using the specialised version (Algorithm 4) of the primal retrieval steps given in [99]. Let us denote x_{ijk} by $x[k : (i, j)]$ for the hypergraph flow variable corresponding to the hyperarc $(\{(i, k), (j, k)\}, (i, j))$. This way the head of

---

Algorithm 3 Algorithm: Value Iteration for LR_λ

INPUT: c_{ij}, cost coefficient for u_{ij} $\forall\, 1 \le i < j \le n$, λ_k, for $k = 4, \ldots, n$.

OUTPUT: $k^*_{(i,j)}$, $t[ij]$ $\forall\, 1 \le i < j \le n$. Dual solution to LR_λ, that is, $\pi_{ij}^{t[ij]}$ $\forall\, 1 \le i < j < k \le n$.

Step 1 $\pi_{ij}^0 \leftarrow M$; $t[ij] \leftarrow 0$; $\forall (i, j) \in E_n$; $t \leftarrow 0$;
Step 2 Iterative Step:
while some π_{ij}^t changed; **do**
  $t \leftarrow t + 1$;
  for $(i, j) \in E_n$ **do**
    $\bar{k} = argmin\{\lambda_k + \pi_{ik}^{t-1} + \pi_{jk}^{t-1} | 4 \le k \le n\}$;
    $\pi_{ij}^t = \min\{\pi_{ij}^{t-1}, \lambda_{\bar{k}} + \pi_{i\bar{k}}^{t-1} + \pi_{j\bar{k}}^{t-1}, c_{ij}\}$;
    if $\pi_{ij}^{t-1} > \pi_{ij}^t$ **then**
      $k^*_{(i,j)} \leftarrow \bar{k}$;
      $t[ij] \leftarrow t$
    end if
  end for
end while
return $k^*_{(i,j)}$, $t[ij]$, $\pi_{ij}^{t[ij]}$ $\forall\, 1 \le i < j \le n$.

---

the hyperarc is seen as (i, j) and the tail set $\{(i, k), (j, k)\}$ corresponds to the insertion of k in (i, j). We wish to retrieve these values for a given dual solution obtained from the value iteration Algorithm 3.

Algorithm 4 Algorithm: Primal Retrieval for LR_λ

INPUT: b_{ij}, demand at node (i, j); $k^*_{(i,j)}$, $t[ij]$ from value iteration algorithm, $\forall\, 1 \le i < j \le n$.

OUTPUT: Primal solution to LR_λ, that is, $x[k : (i, j)]$ $\forall\, 1 \le i < j < k \le n$.

Step 1: Set $x[k : (i, j)] \leftarrow 0$ $\forall (k : (i, j)) \in \mathcal{E}$; create active vertex list $\tilde{V} \leftarrow E_n$; set vertex flows $f_{ij} \leftarrow b_{ij} \ge 0$ $\forall (i, j) \in \tilde{V}$.
Step 2 Iterative Step:
while $\tilde{V} \ne \emptyset$, **do**
  Let $t[(\bar{i}, \bar{j})] = \max\{t[(i, j)] \mid (i, j) \in \tilde{V}\}$.
  Choose the hyperarc with head $(\bar{i}, \bar{j})$ and tail set $\{(\bar{i}, k^*_{\bar{i}\bar{j}}), (\bar{j}, k^*_{\bar{i}\bar{j}})\}$ and
  update $x[k^*_{\bar{i}\bar{j}} : (\bar{i}, \bar{j})] \leftarrow f_{\bar{i}\bar{j}}$;
  $f_{\bar{i},k^*_{\bar{i}\bar{j}}} \leftarrow f_{\bar{i},k^*_{\bar{i}\bar{j}}} + x[k^*_{\bar{i}\bar{j}} : (\bar{i}, \bar{j})]$;
  $f_{\bar{j},k^*_{\bar{i}\bar{j}}} \leftarrow f_{\bar{j},k^*_{\bar{i}\bar{j}}} + x[k^*_{\bar{i}\bar{j}} : (\bar{i}, \bar{j})]$;
  $\tilde{V} \leftarrow \tilde{V} \setminus \{(\bar{i}, \bar{j})\}$.
end while
return $x[k : (i, j)]$ $\forall\, 1 \le i < j < k \le n$.

Thus, we have a way to recover the primal solution from a solution to Problem 8.3 without further increasing the complexity of solving the Problem 8.3. The time complexity of solving this problem using value iteration algorithm and primal retrieval is $O(n^2)$ as the number of rows in the gain free Leontief flow problem (Problem 8.3) is $p_n + (n - 3)$.

8.3.3 Lagrangian Dual of Problem 8.2: LD

As discussed in Sect. 8.2.2 to find the best possible Lagrangian bound, we solve the Langrangian Dual problem. Here we consider the Lagrangian dual given by

Problem 8.4 (*Lagrangian Dual: LD*) Find λ^* real $\ni z(LD) = z(LR_{\lambda^*}) = \max_\lambda \{z(LR_\lambda)\}$.

As mentioned earlier, subgradient optimisation is applied in practice to solve this problem, although we do not have any complexity bounds on using sub-gradient optimisation. Theoretically, we can guarantee a polynomial time algo-rithm to find the optimal value of the Problem 8.2 using Vaidya's sliding objective algorithm [160] as discussed by Bertsimas and Orlin in [31]. Bertsimas and Orlin propose techniques for the solution of the LP relaxation and Lagrangian dual in combinatorial optimisation problems. One can use a simplified and strengthened version as in [3] in place of Vaidya's algorithm. One of the structured LP problems they discuss is $\min\{\sum_{r=1}^N c_r x_r \mid \sum_{r=1}^N A_r x_r = b, x_r \in \mathcal{S}_r, r = 1, \ldots, N\}$. Notice that Problem 8.2.2 can be written in this form using $A^{(k)}$ matrices defined earlier (Definition 4.1) as:

$$\min\{ \sum_{1 \le i < j \le n} c_{ij} w_{ij} \mid \sum_{k=4}^n A^{(k)} x_k + I_{p_n} w = \begin{bmatrix} 1_3 \\ 0 \end{bmatrix}, 1_{p_n} \ge w \ge 0,$$

$$x_k \in \mathcal{S}_k, r = 4, \ldots, n\},$$

with

$$\mathcal{S}_k = \{x_k = (x_{12k}, \ldots, x_{k-2k-1k}) \ge 0 \mid \sum_{1 \le i < j < k} x_{ijk} = 1\}.$$

Let $\{x_k^l, l \in J_k\}$ be the set of extreme pints of $\mathcal{S}_k$. Let $y, \mu \in R^{p_n}$, and $\sigma_k \in R, k = 4, \ldots, n$. We can write the dual of the above problem as:

Problem 8.5

$$z(D) = \max yb + \sum_{1 \le i < j \le n} \mu_{ij} + \sum_{k=4}^n \sigma_k \tag{8.22}$$

subject to $yA^{(k)} x_k^l + \sigma_k \le 0,$ $\qquad\qquad \forall l \in J_k$ and $k \in \{4, \ldots, n\}$, $\tag{8.23}$

$$y + \mu \le c. \tag{8.24}$$

As suggested by Bertsimas and Orlin's approach, we solve the Problem 8.5 using Vaidya's algorithm; which in turn solves a separation problem for all k, given a solution y, μ, σ to this problem. Each separation problem turns out to be a problem of finding the minimum over a multiple choice constrained set having p_k variables. Therefore, it can be done in time $O(p_k)$. So we could apply Theorem 2 of [31] to obtain a bound on the complexity of solving this problem given by

Theorem 8.3 *The optimal solution value of Problem 8.2 and the optimal dual variables can be found in* $O([\sum_{k=4}^{n} p_{k-1}](p_n + (n-3))L + M[p_n + (n-3)](p_n + (n-3))L)$ *arithmetic operations, where L is the input size of the instance, $M[s]$ is the number of arithmetic operations to multiply two matrices of size $s \times s$.* ♡

Following Theorem 1 of [31] the authors discuss finding a solution to the Problem 8.2 without changing the overall complexity of the method. However, we often want to find the optimal solution value rather than the solution of the LP or Lagrangian relaxation, since the solution value can be later used in a branch and bound algorithm. Bertsimas and Orlin in [31] apply their approach to solve Lagrangian Duals discussed in general. So we could apply the same to Problem 8.4. The separation problem in this case turns out to be a Leontief substitution flow problem and so we can solve it in time $O(p_n)$, as observed earlier. Thus using Theorem 4 in [31] we have a polynomial algorithm of complexity $O(p_n(p_n + n - 3)L + M(p_n + n - 3)(p_n + n - 3)L)$ to solve Lagrangian dual LD. Notice that this algorithm complexity-wise is superior to the one mentioned above in Theorem 8.3.

However, in practice, subgradient optimisation and its variants are observed to perform well [114, 167] for solving Lagrangian duals. There are other competing methods like the volume method and bundle method, which could be compared with the subgradient approach to solve Lagrangian dual.

8.4 Hypergraph Flow and MI-Relaxation

A simple trick converts the *MI*-relaxation problem into a minimum-cost hypergraph flow problem. We noticed earlier that the system of equations is not a pre-Leontief system and since there are two $+1$'s in each column; it also does not correspond to a hypergraph flow problem. But multiplying both sides of the multiple choice Eqs. (8.14) by -1, with the rest of the constraints intact, we get a minimum cost hypergraph flow problem corresponding to *MI*-relaxation. However, this destroys the Leontief substitution flow structure as the right-hand side elements are not nonnegative anymore. Here we study the special minimum cost hypergraph flow problem at hand bringing out the changes required in the procedures used in the hypergraph simplex algorithm of [38]. We consider the following hypergraph $\mathcal{H} = (\mathcal{V}, \mathcal{E})$

corresponding to *MI*-formulation. We have, $\mathcal{V} = \{4, \ldots, n\} \cup \{(i, j) \mid 1 \leq i < j \leq n\}$ with $\mathcal{E} = \{(\emptyset, (i, j)) \mid (i, j) \in \mathcal{V}\} \cup_{k=4}^{n} \{(k : (i, j)) \mid 1 \leq i < j < k\}$ where $(k : (i, j))$ denotes the hyperarc $(\{(i, k), (j, k), k\}, (i, j))$, for $1 \leq i < j < k, k \in \mathcal{V}$.

Theorem 8.4 *The hypergraph $\mathcal{H} = (\mathcal{V}, \mathcal{E})$ corresponding to MI-formulation is cycle-free.* ♡

Proof Vertices in $S = \{4, \ldots, n\}$ have no hyperarcs entering any of $k \in S$. Thus any directed path starting from k cannot end in k. Therefore, there are no cycles involving $k \in S$. Consider any $(i, j) \in \mathcal{V}$, for any $1 \leq i < j \leq n$.

Case 1: $(i, j) \in \mathcal{V}, 1 \leq i < j \leq 3$. Since none of these vertices is in the tail set of any hyperarc, a directed cycle involving such an (i, j) is not possible. Case 2: $(i, j) \in \mathcal{V}, 1 \leq i < j, 4 \leq j \leq n$. Suppose for some (i_0, j_0) there is a directed cycle, then (i_0, j_0) is the head for a hyperarc e and is in the tail set of another hyperarc, e'. Therefore, e' has to be an arc $(j_0 : (u, v))$ for some $u < v < j_0$ with u or $v = i_0$ and e is either $(\emptyset, (i_0, j_0))$ or $(r : (i_0, j_0))$ for some $n \geq r > j_0$.

First, we show that $e = (\emptyset, (i_0, j_0))$ is not possible. In any directed path, if e appears as e_i for some $1 \leq i \leq q$ then the vertex v_i is required to belong to $\{h_{e_{i-1}}\} \cup T_{e_i}$. But $T_{e_i} = T_e = \emptyset$ implies e cannot appear in any such directed path.

So $e = (r : (i_0, j_0))$ for some $n \geq r > j_0$. Now we have in the directed path,

$$\ldots, (r : (i_0, j_0)), (i_0, j_0), (j_0 : (u, v)), \ldots$$

with $u < v < j_0$ and one of u or $v = i_0$.

Thus any directed path $P_{st} = (v_1 = s = (i, j), e_1, v_2, \ldots, e_q, v_{q+1} = t)$ in $\mathcal{H}$ is such that $h_{e_q} = t$ and $t = (a, b)$ with $max\{a, b\} < j$. Therefore $t = s$ is not possible, and hence $\mathcal{H}$ is cycle-free. □

Theorem 8.5 *Given any pedigree P, we have a spanning hypertree of $\mathcal{H} = (\mathcal{V}, \mathcal{E})$ given by $(\mathcal{V}, E_T)$ with*

$$E_T = \{(k : (i, j)) \mid (i, j) \in E(P)\} \cup \{(\emptyset, (i, j)) \mid (i, j) \in E_{n-1} \setminus E(P)\},$$

where, $E(P) = \{(i_k, j_k) \mid 4 \leq k \leq n, \ni P = ((i_4, j_4), \ldots, (i_n, j_n))\}$ is the given pedigree. ♡

Proof Since $E_T \subset \mathcal{E}$ and $(\mathcal{V}, E_T)$ is a subhypergraph of $\mathcal{H}$, $(\mathcal{V}, E_T)$ is cycle-free. Let $R = \{k \mid 4 \leq k \leq n\}$, then $N = \mathcal{V} \setminus R$. So $(\mathcal{V}, E_T)$ satisfies the requirement $R \cap N = \emptyset$. No hyperarc in E_T has a vertex in R as its head. Every vertex $v = (i, j) \in N$

either has a unique hyperarc $(k : (i, j))$ entering (i, j) if $(i, j) \in E(P)$ or has a unique hyperarc $(\emptyset, (i, j))$ entering (i, j) if $(i, j) \notin E(P)$. Thus $(\mathcal{V}, E_T)$ is indeed a hypertree. Notice that it is a spanning hypertree as well, as all vertices in $\mathcal{V}$ are spanned.

Since we have $m = |\mathcal{V}|$ is the number of vertices and we have only $|E_T|$ hyperarcs, we need to add $n - 3$ other hyperarcs such that the incidence matrix corresponding to the spanning hypertree is extended to a basis of size $m \times m$. The $n - 3$ hyperarcs external to the hyperarcs in E_T are such that they have at least one non-zero element among the last E_n elements of the column. Let $\mathcal{T}_R$ be a traverse of $(\mathcal{V}, E_T)$ given by

$$(R, (\emptyset, (i_n, n)), (i_n, n), (\emptyset, (j_n, n)), (j_n, n), (n : i_n, j_n), \ldots,$$

$$(4 : i_4, j_4), (\emptyset, (i, j)), (i, j), \forall (i, j) \notin E(P)).$$

This introduces an order among the vertices and hyperarcs and this order is used in rearranging the incidence matrix of the spanning hypertree.

Remark

1. Notice that $\forall e \in \mathcal{E} \setminus E_T$, $T_e \cup \{h_e\}$ is not a subset of R as every hyperarc not in E_T has $h_e \in E_n$ and so not in R.
2. In general hypergraph flow problems, we may not always have an initial primal feasible solution to the hypergraph flow problem. But for the *MI*-hypergraph flow problem, we have a feasible basis given by the spanning hypertree corresponding to any pedigree, so we can start the hypergraph simplex algorithms without phase I using artificial variables.
3. The set of linearly independent columns corresponding to the hyperarcs in E_T, needs to be expanded to a basis of size $m \times m$.

Example 8.1 Consider $n = 6$ and the pedigree in $\mathcal{P}_6$ given by $P = ((1, 3), (1, 4), (2, 3))$. The corresponding spanning tree is given in Fig. 8.3. The corresponding traverse $\mathcal{T}_R$ of the spanning hypertree with root $R = \{4, 5, 6\}$ and $E_T = \{(4 : (1, 3)), (5 : (1, 4)), (6 : (2, 3))\} \cup \{(\emptyset, (i, j)) \mid (i, j) \in E_5 \text{ and } (i, j) \notin \text{the pedigree } P\}$ is:

$\mathcal{T}_R = (R, (\emptyset, (2, 6)), (2, 6), (\emptyset, (3, 6)), (3, 6), (6 : (2, 3)), (2, 3), (\emptyset, (1, 5)), (1, 5), (\emptyset, (4, 5)),$

$(4, 5), (5 : (1, 4)), (1, 4), (\emptyset, (3, 4)), (3, 4), (4 : (1, 3)), (1, 3), (\emptyset, (1, 2)), (1, 2), (\emptyset, (2, 4)), (2, 4),$

$(\emptyset, (2, 5)), (2, 5), (\emptyset, (3, 5)), (3, 5), (\emptyset, (1, 6)), (1, 6), (\emptyset, (4, 6)), (4, 6), (\emptyset, (5, 6)), (5, 6)).$

We expand this to a basis by adding $3 (= 6 - 3)$ more external hyperarcs $(4 : (1, 2), (5 : (3, 4) \text{ and } (6 : (3, 5))$ that are linearly independent of the set of columns corresponding to the 15 hyperarcs in E_T. This initial basis is shown in Table 8.1. Notice that it is an upper triangular matrix.

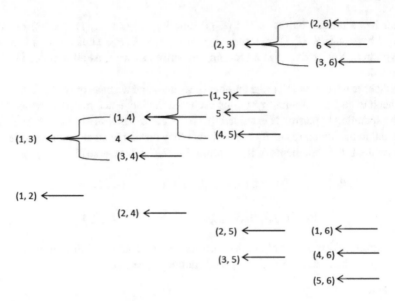

Fig. 8.3 Spanning tree for Example 8.1

We briefly reproduce the main steps of the Hypergraph Simplex from [38]. Since we have an initial basis, we proceed to discuss the hypergraph simplex method for solving the minimum cost hypergraph flow problem. Each basis M corresponds to a pair $(\mathcal{T}_R, E_X)$ where $\mathcal{T}_R$ is a spanning hypertree of the sub-hypergraph corresponding to M. E_X is the set of the external hyperarcs, that is, the basic hyperarcs other than the hyperarcs in the spanning hypertree. The number of external arcs is the same as the number of root vertices of $\mathcal{T}_R$. The root matrix M_R is given by rows corresponding to the root vertices and columns corresponding to the external hyperarcs. M corresponds to a basis if and only if the matrix M_R is non-singular. M_R can be completely characterised in terms of hypergraph concepts like flows and potentials. Let $\bar{b} = (\bar{b}(R), \bar{b}(N))$ be the demand vector induced on the vertices by the flows on the non-basic hyperarcs. Then the system $Mf = \bar{b}$ can be solved by the procedure PRIMAL, and $\pi M = \bar{c}$ can be solved by the procedure DUAL given in [38]. The optimality of the current basis is checked using the reduced cost and thus computed. The optimality conditions are, as usual, based on the reduced costs: non-basic hyperarcs must have reduced costs ≥ 0 if their flow is zero, and reduced costs ≤ 0 if their flow is at the upper bound. If these conditions are satisfied, M is optimal and the algorithm terminates. The steps of the simplex method are similarly inherited in hypergraph simplex, like choosing the hyperarc to enter and choosing the hyperarc to leave the basis; but these are done on the hypergraph, using hypergraph concepts.

Table 8.1 Initial basis for Example 8.1. This corresponds to the spanning hypertree shown in Fig. 8.3 (zeros are suppressed)

hyperarc->	06		05			04										04	05	06
vertex	26	36	23	15	45	14	34	13	12	24	25	35	16	46	56	12	34	35
4							−1									−1		
5				−1													−1	
6		−1																−1
26	1	−1																
36	1	−1																−1
23	1																	
15				1	−1													
45					1	−1							−1					
14						1	−1						−1					
34							1	−1									1	
13								1										
12									1							1		
24										1							−1	
25											1							
35												1					−1	1
16													1					
46														1				
56															1			−1

Since the reported performance of the hypergraph simplex compared to that of generic LP solvers is superior, it is worth specialising the procedures FLOW,[2] POTENTIAL, PRIMAL, and DUAL of the hypergraph simplex algorithm to the particular special hypergraph corresponding to *MI*-relaxation. Once these are implemented in a computer, the new algorithms based on the value iteration algorithm for the Leontief substitution flow problem and the special minimum cost hypergraph flow algorithm could be compared with commercially available/open-source LP solvers.

A four-phase approach to experimentally compare the efficacy of the suggested new algorithms is planned. Phase 1 is generating the computer codes for the specialised subroutines and algorithms and testing them. Phase 2 is the implementation of a prototype of the algorithms in phase 1. Phase 3 consists of optimising the prototype from phase 2. Phase 4 consists of computational experiments. The first phase in this plan appears in [13]. Next, we present some computational experience with *MI*-formulation using the branch-and-bound approach, a cleaver method of exhaustion that is used in practice.

---

[2] The specialised version of the procedure FLOW from [38] is included in the appendices, see Appendix 3 given in Sect. 8.8.3.

8.5   Clever Methods of Exhaustion: Relaxations for the TSP

One successful variant of the method of exhaustion is the cleaver combination of partitioning the solution set—branching and terminating or postponing the exhaustion of a partition using a bound on the objective values in that partition or pruning the branch so no further search is done by using cuts. We have already mentioned in Chap. 1 the tremendous success of branch-and-cut by *Concorde* software. We review some of the branching strategies used in the literature for solving *TSP* instances.

For the branch and bound method to perform well, it is important to start with an LP relaxation of the IP problem with a small gap. Different relaxations of the *TSP* have been considered by researchers to apply in the branch and bound methods. We give some of these results below.

The assignment problem (AP) relaxation   of the *ATSP* is used by Eastman [62], Little et al. [113], and Bellmore and Malone [29]. The AP is given by the objective function of the *DFJ* formulation, and constraints (8.25)–(8.28).

$$\sum_{(i,j)\in A} x_{ij} = 1, \ \forall i \in V, \tag{8.25}$$

$$\sum_{(i,j)\in A} x_{ij} = 1, \ \forall j \in V, \tag{8.26}$$

$$x_{ij} \leq 1, \ \forall i, j \in V, \tag{8.27}$$

$$x_{ij} \geq 0, \ \forall i, j \in V. \tag{8.28}$$

The solution to the AP is either a directed tour or a set of directed subtours. Eastman used the network flow algorithm by Ford and Fulkerson [72] to solve the AP. The AP can be solved using the Hungarian method in $O(n^3)$ time; there are other algorithms for this problem with better computational efficiency [2].

The 2-matching relaxation   is used by Bellmore and Malone [29] for the *STSP*. The answer to the 2-matching problem is either a tour or a collection of subtours. The 2-matching problem has the same objective function as the *DFJ* formulation subject to 2-matching constraints of *DFJ* formulation, (8.27), and (8.28).

1-tree relaxation   of the *STSP* was first used by Held and Karp [93] and Christofides [43]. Let V' be $V - \{1\}$. This relaxation is given by constraints (8.27), (8.28), and constraints (8.29)–(8.31).

$$\sum_{(i,j)\in S\times(V'-S),j>i} x_{ij} + \sum_{(i,j)\in(V'-S)\times S,j>i} x_{ij} \geq 1, \ \forall S \subset V', \tag{8.29}$$

$$\sum_{i\in V}\sum_{j>i} x_{ij} = n, \tag{8.30}$$

$$\sum_{i\in V} x_{1j} = 2, \tag{8.31}$$

The n-path problem   is about finding the shortest path in a graph that includes n nodes (n-path) starting and ending at some node $v \in V$. The LP relaxation of the n-path problem is first used by Houck et al. [95] for the *TSP*. The n-path problem can be solved using dynamic programming in $O(n^3)$ steps [24].

The LP with cutting planes   was first suggested by Gomory [79] for solving generic integer programming optimisation problems. Crowder and Padberg [54] used this solution method for the *STSP*. The main feature of their method is finding appropriate inequalities to use as cutting planes in each step [24].

8.5.1 Branching Methods

Borrowing the notations used by Balas and Toth [24], for defining the sub-problems in a branching tree. Starting with the root problem labelled as Problem 1, we use string labels for the sub-problems in a way that they show the hierarchy of the problem in the branch and bound tree. Given some problem m in the branching tree, let $\mathcal{E}_m$ indicate the set of edges (i, j) that are excluded from problem m, and let $\mathcal{I}_m$ indicate the set of edges included in the problem. Using variables x_{ij}, the sets $\mathcal{E}_m$ and $\mathcal{I}_m$ for some sub-problem m can be defined by the following condition.

$$\begin{cases} (i, j) \in \mathcal{I}_m, \text{ if } x_{ij} = 1 \text{ in problem } m, \\ (i, j) \in \mathcal{E}_m, \text{ if } x_{ij} = 0 \text{ in problem } m. \end{cases} \tag{8.32}$$

The solution to problem m is given subject to the constraints of the predecessor problem of m and subject to Condition (8.32). Given problem m, let the rth successor of problem m be labelled as problem mr. We use the notations $\mathcal{I}_{mr}$ and $\mathcal{E}_{mr}$ to indicate the set of variables that are included in and excluded from the mr problem respectively.

Many different branching rules have been suggested for *TSP*. Little et al. [113] designed a branching rule that creates two sub-problems where some arc (i, j) is included in one branch and excluded from the other branch. This rule does not use the structure in the *TSP* and can be applied to any IP problem. Little et al. do not solve the LP at each node but they calculate a lower bound for the LP using the dual solution. Their algorithm has more branching steps but requires less computation at each step compared to other methods [110]. At each branching step, Little et al. [113] select a pair (i, j) such that the sub problem corresponding to $x_{ij} = 0$ would give a bound as large as possible [110]. The branching rule can be expressed as follows.

$$\begin{cases} \mathcal{E}_{m1} = \mathcal{E}_m \cup \{(i, j)\}, \ \mathcal{I}_{m1} = \mathcal{I}_m, \\ \mathcal{E}_{m2} = \mathcal{E}_m, \qquad\qquad \mathcal{I}_{m2} = \mathcal{I}_m \cup \{(i, j)\}. \end{cases} \tag{8.33}$$

For more branching rules and *TSP* relaxations we refer the reader to the works by Balas and Toth [24], and Applegate et al. [5].

8.6   Branch and Bound Method for MI Formulation

Similar to the notations used in the previous section for x_{ij} variables, we use the
following notations based on the *MI*-formulation structure to define subproblems
in a branch and bound tree. Let the sets $\mathcal{E}_m$ and $\mathcal{I}_m$ correspond to the sets of x_{ijk}
variables that are excluded from or included in the solution of some subproblem m
respectively. We define $\mathcal{E}_m$ and $\mathcal{I}_m$ by the following condition for the subproblem m.

$$\begin{cases} (i, j, k) \in \mathcal{I}_m, & \text{if } x_{ijk} = 1, \\ (i, j, k) \in \mathcal{E}_m, & \text{if } x_{ijk} = 0. \end{cases} \tag{8.34}$$

We can use sets $\mathcal{E}_m$ and $\mathcal{I}_m$ to define subproblems $m1$ and $m2$ that partition the
solution space of problem m into two regions.

Example 8.2 Consider this solution to a 13-city *MI*-relaxation problem, $x_{2,3,4} =$
1, $x_{1,2,5} = 1$, $x_{1,3,6} = 1$, $x_{2,5,7} = 1/2$, $x_{2,3,4} = 1$, $x_{2,3,4} = 1$, $x_{2,3,4} = 1$, $x_{2,3,4} = 1$,
$x_{2,3,4} = 1$, $x_{2,3,4} = 1$, $x_{2,3,4} = 1$, $x_{2,3,4} = 1$, $x_{3,6,7} = 1/2$, $x_{3,7,8} = 1/2$, $x_{6,7,8} = 1/2$,
$x_{2,4,9} = 1$, $x_{3,6,10} = 1/2$, $x_{6,8,10} = 1/2$, $x_{7,8,11} = 1$, $x_{3,8,12} = 1/2$, $x_{3,10,12} = 1/2$, and
$x_{6,10,13} = 1$. Having $x_{3,8,12} = 0.5$, and $x_{3,10,12} = 0.5$, suggests that edge (3, 12) is created by having variables $x_{3,j,12}$ to take values greater than zero. We use the following
branching rule that partitions the solution space into two regions $\sum_{i \neq 3, i < j < 12} x_{i,j,12}$,
and $\sum_{i=3, i < j < 12} x_{i,j,12}$. Similarly, for $k = 8$ we have $x_{3,7,8} = 0.5$, and $x_{6,7,8} = 0.5$,
so we can have $\sum_{j \neq 7} x_{ij8} = 0$, and $\sum_{j=7} x_{ij8} = 0$. And finally, based on having
$x_{3,6,10} = 0.5$ and $x_{6,8,10} = 0.5$, we can use $\sum_{i \neq 6 \wedge j \neq 6} x_{i,j,10} = 0$, and
$\sum_{i=6 \vee j=6} x_{i,j,10} = 0$, to partition the solution space into two regions.

The following branching rules generalise the branching formulations given in
Example 8.2. These rules partition the solution space into at least two regions:

MI Branching Rule 1 (MIR₁)
Given the solution to some problem m, if for some $\hat{k}$ and some $\hat{i}$ we have $0 < x_{\hat{i} j \hat{k}} < 1$,
then the successors of m are partitioned into two groups based on the following rules.

$$\begin{cases} \mathcal{E}_{m1} = \mathcal{E}_m \cup \{(i, j, k) | i = \hat{i}, k = \hat{k}, i < j < k\}, \mathcal{I}_{m1} = \mathcal{I}_m, \\ \mathcal{E}_{m2} = \mathcal{E}_m, I_{m2} = \mathcal{I}_m \cup \{(i, j, k) | i = \hat{i}, k = \hat{k}, i < j < k\}, \end{cases} \tag{8.35}$$

Similarly, if for some $\hat{k}$ and some $\hat{j}$ for some problem m we have $0 < x_{i\hat{j}\hat{k}} < 1$,
then the successors of m are partitioned into two groups using the following rules.

$$\begin{cases} \mathcal{E}_{m1} = \mathcal{E}_m \cup \{(i, j, k) | j = \hat{j}, k = \hat{k}, 1 < i < j\}, \mathcal{I}_{m1} = \mathcal{I}_m, \\ \mathcal{E}_{m2} = \mathcal{E}_m, \mathcal{I}_{m2} = \mathcal{I}_m \cup \{(i, j, k) | j = \hat{j}, k = \hat{k}, 1 < i < j\}, \end{cases} \tag{8.36}$$

MI **Branching Rule 2 (MIR$_2$)**

Given some problem m, if for some $\hat{k}$ and some $\hat{i}$, we have $0 < x_{\hat{i}_1 \hat{j} k}, x_{\hat{i}_2 \hat{j} k} < 1$, the successors of m can be defined as follows.

$$\begin{cases} \mathcal{E}_{m1} = \mathcal{E}_m, \mathcal{I}_{m1} = \mathcal{I}_m \cup \{(\hat{i}_1 \hat{j}\hat{k})\}. \\ \mathcal{E}_{m2} = \mathcal{E}_m, \mathcal{I}_{m2} = \mathcal{I}_m \cup \{(\hat{i}_2 \hat{j}\hat{k})\}. \\ \mathcal{E}_{m3} = \mathcal{E}_m \cup \{(\hat{i}_1 \hat{j}\hat{k}), (\hat{i}_2 \hat{j}\hat{k})\}, \mathcal{I}_{m3} = \mathcal{I}_m. \end{cases} \tag{8.37}$$

Similarly, for some $\hat{k}$ and some $\hat{i}$, we have $0 < x_{\hat{i} \hat{j}_1 k} < 1$, and $0 < x_{\hat{i} \hat{j}_2 k} < 1$, the successors of m can be defined as follows.

$$\begin{cases} \mathcal{E}_{m1} = \mathcal{E}_m, \mathcal{I}_{m1} = \mathcal{I}_m \cup \{(\hat{i} \hat{j}_1 \hat{k})\} \\ \mathcal{E}_{m2} = \mathcal{E}_m, \mathcal{I}_{m2} = \mathcal{I}_m \cup \{(\hat{i} \hat{j}_2 \hat{k})\} \\ \mathcal{E}_{m3} = \mathcal{E}_m \cup \{(\hat{i} \hat{j}_1 \hat{k}), (\hat{i} \hat{j}_2 \hat{k})\}, \mathcal{I}_{m3} = \mathcal{I}_m. \end{cases} \tag{8.38}$$

MI **Branching Rule 3 (MIR$_3$)**

This branching rule is the same as the generic branching rule for the IP methods. Given an *MI*-relaxation solution for some $0 < x_{\hat{i} \hat{j} \hat{k}} < 1$, we define the following branching rule.

$$\begin{cases} \mathcal{E}_{m1} = \mathcal{E}_m, \mathcal{I}_{m1} = \mathcal{I}_m \cup \{(\hat{i}, \hat{j}, \hat{k})\}, \\ \mathcal{E}_{m2} = \mathcal{E}_m \cup \{(\hat{i}, \hat{j}, \hat{k})\}, \mathcal{I}_{m2} = \mathcal{I}_m, \end{cases} \tag{8.39}$$

When choosing x_{ijk} variables to branch on, we have the option of choosing variables with k values as close to n or as close to 4 as possible. For example, in the solution given in Example 8.2, we can choose between branching on x_{ijk} variables with $k = 7$ or $k = 12$. We represent the combination of a branching rule MIR$_i$ with branching on variables of either the greatest value of k or smallest values of k, using MIR$_{i,1}$, and MIR$_{i,2}$, respectively.

In Sect. 8.8.1 we give the comparative computational results on these different branching rules for some TSPLIB [147] instances.

8.7 Conclusion

We would have come full circle if we go back to the definition of a pedigree given in Chap. 1, namely, Definition 1.5.

Definition 8.11 (*Pedigree*) A spanning set P in D is called a *Pedigree*, if and only if it is a tree,

1. rooted at $\{1, 2, 3\}$,
2. has exactly one element from each $\Delta^k, k = 3, \ldots, n$, so $|P| = n - 2$, and
3. every element of P other than the root has a generator in P, and the common edges are all distinct. ♣

Though the other definitions of pedigree given in Chaps. 3 and 4 were used in proving the results in the book, there is a reason for the definition of a pedigree using triangles.

Firstly, we repeat some definitions from algebraic topology (See [124], for instance).

Let an arbitrary fixed field be denoted by $\mathbb{F}$.

Definition 8.12 A d-dimensional simplex, abbreviated as d-simplex, is an oriented set $\sigma \subseteq [n]$ with $|\sigma| = d + 1$. Here, orientation is expressed by viewing σ as an ordered $(d + 1)$-tuple $\sigma = (s_1, s_2, \ldots, s_{d+1})$, where $s_1 < s_2 < \cdots < s_{d+1}$.

A face of a σ is any (oriented) simplex supported on the subset of $V[\sigma] = \{s_1, s_2, \ldots, s_{d+1}\}$. ♣

Triangles are 2-simplices. Now let us consider a simplicial complex as defined below:

Definition 8.13 A simplicial complex K is a collection of simplices over $[n]$ closed under containment, i.e., if $\sigma \in K$, then so are all the faces of σ. As before, $\sigma \in K$ is called a face of K. The dimension of K is the largest dimension over all its faces. Such maximal faces are called facets. If all maximal faces of K are of the same dimension, d, we say K is a pure d-dimensional complex. ♣

The complete d-dimensional complex $K_n^d = \{\sigma \subset [n] \,|\, |\sigma| \leq d + 1\}$ contains all possible simplices over $[n]$ of dimension at most d. Let K be a d-complex. We denote by $K^{(d)}$ the set of all d-faces of K. A complex K is said to have full d-skeleton if $K^{(d)}$ contains all $\binom{n}{d+1}$ d-simplices.

Example 8.3 Consider $K = K_n^2$, the complete 2-dimensional complex containing all possible simplices over $[n]$ of dimension at most 2. Then Δ, the collection of all triangles, is K^2. Therefore, K has full 2-skeleton.

Definition 8.14 A d-chain of K is formal sum $C_d = \sum_{\sigma_i \in K^{(d)}} c_i \sigma_i$ with $c_i \in \mathbb{F}$. Alternatively, C_d can be viewed as a $|K^{(d)}|$-dimensional $\mathbb{F}$-valued vector indexed by members of $K^{(d)}$. ♣

Example 8.4 Given $n = 5$, consider K a 2-simplicial complex given by

$$K = \{(1, 2, 3), (2, 3, 4), (2, 4, 5), (1, 2, 5)\}.$$

A 2-chain of K is given by $(1, 2, 3) - (2, 3, 4) + (2, 4, 5)$.

> **Definition 8.15** For a d-simplex $\sigma = (s_1, s_2, \ldots, s_{d+1})$, its d-boundary is defined as a $(d - 1)$ chain , $\partial_d(\sigma) = \sum_{i=1}^{d+1} (-1)^{i-1} (\sigma \backslash s_i)$, where $(\sigma \backslash s_i)$ is an oriented facet of σ obtained by the deletion of s_i. ♣

Example 8.5 For a 2-simplex $\sigma = (2, 4, 5)$, its 2-boundary is given by, $\partial_2(\sigma) = (4, 5) - (2, 5) + (2, 4)$. This results in the cycle $4- > 5- > 2- > 4$, notice that $-(2, 5)$ reverses the orientation.

With these definitions, we can see how a pedigree is a pure 2-dimensional simplicial complex, with some additional restrictions; and a boundary of a pedigree is a Hamiltonian cycle. The question then is, how beneficial algebraic topology would be in studying combinatorial optimisation problems?

L. J. Billera [32] provides an early expository account describing the use of methods of commutative algebra to solve problems concerning the enumeration of faces of convex polytopes. There we also have a glimpse of simplicial complexes and the application of some of these methods to the study of integer solutions to systems of linear inequalities.

> ...begin to see a striking connection between the numbers of faces of a polytope and the integer points in a related system of convex cones. The possibility of using these methods to shed light on general integer programming problems is one of the most exciting aspects of this area of research.

Thus we have the research question: "How can algebraic topology enhance our understanding of combinatorial optimisation problems?"

The use of hypergraph simplex suggested in solving the *MI*-relaxation problem, Lagrangian relaxation approaches and other computational experiences outlined in this chapter indicate opportunities for designing novel algorithms for solving *STSP* from this new perspective. This provides another direction of research in solving *COPs*.

We have come a long way since the time Dantzig, Fulkerson and Johnson gave the standard formulation for *STSP*, and today we can solve enormously huge problem instances using the *Concorde TSP* Solver. But showing *STSP* can be solved efficiently, an unexpected theorem proved in this book, is expected to open up renewed interest in solving difficult combinatorial problems. Also, it might challenge assumed computer security paradigms. It may not be a tall claim to conclude the book with the following hopeful note:

The future is bright despite the complexities we face.

8.8 Appendices

The following three subsections give the results of the computational experiments
mentioned in this chapter. So, refer to different sections of Chap. 8 for the nomen-
clature used to refer to different sets of instances, formulations, and branching rules.

8.8.1 Appendix 1: On Branch & Bound Computational Comparisons

We applied different *MI* branching rules, $MIR_{i,j}$ for $i = 1, 2, 3$ and $j = 1, 2$, on
some TSPLIB instances. We used Cplex 9.1 for solving the subproblems in each tree
node. The solution times are compared in Table 8.2 and the sizes of the branch and
bound trees are compared in Table 8.3. These results are also illustrated in Fig. 8.4.
The median, minimum, and maximum values of the solution times and the size of
the trees are shown in this figure. The first and third quartiles are illustrated using
boxes. From Fig. 8.4 we can observe that $MIR_{1,1}$ and $MIR_{1,2}$ provide smaller branch
and bound trees compared to other methods. The solution times for $MIR_{1,1}$ are less
than those for other rules, except for problem *st*70, a Euclidean 70-city problem.
For the MIR_1 rules, branching on the variables with the smallest value of k seems to
perform better than using larger values for k.

We applied branch and bound with the *DFJ* [56], Wong [165], Claus [46], and
Carr [40] formulations and compared with branch and bound on the *MI*-formulation.
Robert Carr's algorithm uses a relaxation of the *TSP* called *cycle-shrink*.

Table 8.2 CPU seconds for branch and bound method with *MI* branching rules

Problem	$MIR_{1,1}$	$MIR_{2,1}$	$MIR_{3,1}$	$MIR_{1,2}$	$MIR_{2,2}$	$MIR_{3,2}$
bayg29	1.13	1.41	1.04	**0.75**	1.97	1.97
bays29	1.47	1.43	2.23	**1.00**	3.23	3.29
dantzig42	5.44	3.27	6.07	**2.34**	3.38	5.22
swiss42	**1.42**	1.83	1.45	1.66	3.48	2.40
att48	**4.70**	8.69	8.46	16.89	24.52	17.18
hk48	2.93	3.64	**2.92**	9.31	12.70	12.53
brazil58	**6.23**	8.34	6.98	6.65	9.09	6.77
st70	408.91	588.33	629.31	98.46	**59.38**	206.07
eil76	187.67	504.81	**139.58**	355.28	408.42	851.87
rd100	**305.99**	441.49	361.57	394.76	652.46	444.35
eil101	**1209.30**	3342.42	3956.10	2254.70	3618.60	3693.50
lin105	**113.78**	148.27	119.61	187.84	538.37	384.00
gr120	$>10^4$	$>10^4$	$>10^4$	–	–	–

Table 8.3 Size of the branch and bound tree

Problem	Carr	Claus	DFJ	MIR$_{1,1}$	MIR$_{2,1}$	MIR$_{3,1}$	MIR$_{1,2}$	MIR$_{2,2}$	MIR$_{3,2}$	Wong
bayg29	7	19	5	5	7	5	5	**3**	10	19
bays29	11	33	7	7	7	11	11	**5**	16	33
dantzig42	11	55	5	11	7	13	13	5	7	49
swiss42	5	17	5	**3**	4	**3**	**3**	7	5	27
att48	9	25	7	**5**	10	9	17	25	17	27
hk48	7	7	7	**3**	4	**3**	9	13	13	13
brazil58	3	25	7	**3**	4	**3**	**3**	4	**3**	21
st70	–	–	117	99	147	151	19	**13**	39	–
eil76	–	–	11	19	53	**13**	33	41	73	–
rd100	–	–	7	**7**	11	9	9	13	9	–
eil101	–	–	47	**19**	70	81	35	62	63	–
lin105	–	–	**3**	**3**	4	**3**	5	13	9	–
gr120	–	–	2055	**71**	90	93	–	–	–	–

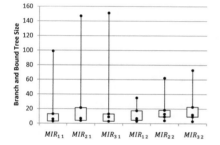 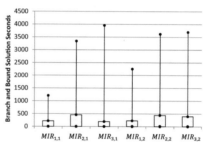

Fig. 8.4 Branch and bound results-*MI* branching rules. Solution seconds (left) and tree size (right)

Cycle-shrink is a compact description of the subtour elimination polytope. In [19] we show the equivalence of *MI*-formulation and cycle-shrink formulation. So this computational comparison is interesting. The results are given in Tables 8.3 and 8.4, and illustrated in Fig. 8.5. The number of violated subtour elimination constraints for the *DFJ* formulation that are found and added to the subproblems is shown in Table 8.4 in column *SEC*. Apart from the *DFJ* formulation, all three *MI* branching rules provide the smallest size for branching trees and require the least amount of computational time. For the *DFJ* formulation, the size of the branch and bound tree is greater than that of *MI*, except for problem *eil*76, but the computational time for the *DFJ* formulation is less than that for the *MI*-formulation. This is probably due to the small size of the LPs solved at each branch and bound node. The size of the branch and bound tree is significantly larger for the *DFJ* formulation for problem gr120, which is a 120-city problem with geographical distance, compared to the *MI*-formulation. The *MI* branching methods provide smaller branch and bound trees for gr120, although their solution time is greater than that of the *DFJ* formulation.

Table 8.4 Solution seconds for branch and bound on different methods

Problem	Carr	Claus	DFJ	SEC	MIR$_{1,1}$	MIR$_{2,1}$	MIR$_{3,1}$	Wong
bayg29	29.87	136.07	3.97	65	1.15	1.41	**1.04**	558.0
bays29	45.88	218.88	2.26	34	1.47	**1.43**	2.23	934.7
dantzig42	547.43	4706.28	3.37	32	5.44	**3.27**	6.07	13777.5
swiss42	260.33	1418.20	2.85	44	**1.42**	1.83	1.45	4878.5
att48	3097.80	6765.88	9.26	285	**4.70**	8.69	8.46	15277.8
hk48	1964.80	2415.82	4.84	139	2.93	3.64	**2.92**	6764.8
brazil58	5075.50	18168.93	6.53	131	**6.23**	8.34	6.98	23619.5
st70	–	–	**125.21**	90757	408.91	588.33	629.31	–
eil76	–	–	**12.70**	106	187.67	504.81	139.58	–
rd100	–	–	**43.15**	456	305.99	441.49	361.57	–
eil101	–	–	**174.90**	14006	1209.30	3342.42	3956.10	–
lin105	–	–	**46.17**	138	113.78	148.27	119.61	–
gr120	–	–	**5411.50**	4930140200	$>10^4$	$>10^4$	$>10^4$	–

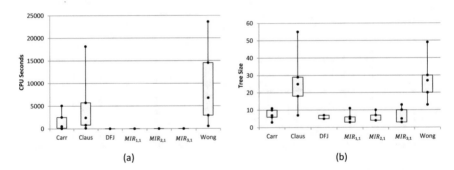

(a)   (b)

Fig. 8.5 Branch and bound results-TSP formulations. **a** Solution seconds and **b** Tree size

8.8.2 Appendix 2: On Computational Comparisons

In this section we provide excerpts from Ardekani's doctoral thesis [90]: In Table 6.1 of [90] (reproduced here as Fig. 8.6) we have the computational comparisons of *DFJ* relaxation and *MI*-relaxation on *STSP* instances from TSPLIB [147]. Here, column 1 gives the problem name. The values of the 2-matching relaxation and its gap are shown in columns LP1 Value and LP1 Gap respectively. The number of violated subtour elimination constraints found using two different algorithms are shown in columns SEC1 and SEC2, respectively. The solutions given by *MI*-formulation are given in the table as well as the CPU seconds for the *MI* and the *DFJ* formulations. The results for the *MI*-relaxation formulation are given separately in Table 6.2 of [90] (reproduced here as Figs. 8.7 and 8.8) in more detail. Table 6.2 includes the optimal solution of the instances and the optimal solution of the MI-relaxation. The solutions

Table 6.1: Results for the DFJ Formulation Compared with the MI-relaxation

Problem	Optimal Value	MI Value	LP$_1$ Value	MI gap	LP$_1$ gap	SEC$_1$	SEC$_2$	CPU Seconds DFJ	MI
gr17	2085	2085.0	1684.0	0.00%	19.23%	11	0	1.45	**0.02**
gr21	2707	2707.0	2707.0	0.00%	0.00%	0	0	0.14	**0.02**
gr24	1272	1272.0	1224.5	0.00%	3.73%	2	2	0.73	**0.03**
fri26	937	937.0	880.0	0.00%	6.08%	14	2	2.16	**0.02**
bayg29	1610	1608.0	1546.0	0.12%	3.98%	8	2	1.40	**0.03**
bays29	2020	2013.5	1944.0	0.32%	3.76%	3	1	0.63	**0.03**
dantzig42	699	697.0	641.0	0.29%	8.30%	5	2	1.24	**0.14**
swiss42	1273	1272.0	1214.5	0.08%	4.60%	3	8	2.03	**0.16**
att48	10628	10604.0	10041.5	0.23%	5.52%	27	7	6.10	**0.58**
gr48	5046	4959.0	4769.0	1.72%	5.49%	5	4	1.80	**0.69**
hk48	11461	11444.5	11197.0	0.14%	2.30%	14	5	3.30	**0.52**
eil51	426	422.5	416.5	0.82%	2.23%	4	0	**0.74**	0.88
berlin52	7542	7542.0	7163.0	0.00%	5.03%	6	0	1.05	**0.97**
brazil58	25395	25345.5	20896.0	0.19%	17.72%	9	12	4.73	**1.42**
st70	675	671.0	623.5	0.59%	7.63%	19	16	10.08	**3.39**
eil76	538	537.0	534.0	0.19%	0.74%	4	0	**0.87**	9.92
pr76	108159	105120.0	98994.5	2.81%	8.47%	5	5	3.16	**2.80**
rat99	1211	1206.0	1198.0	0.41%	1.07%	9	11	**12.13**	14.25
kroA100	21282	20936.5	19378.5	1.62%	8.94%	18	23	**22.79**	29.89
kroB100	22141	21834.0	20339.5	1.39%	8.14%	14	25	**23.09**	25.25
kroC100	20749	20472.5	19705.0	1.33%	5.03%	27	18	23.20	**21.88**
kroD100	21294	21141.5	19952.5	0.72%	6.30%	23	25	27.87	**27.78**
kroE100	22068	21799.5	20622.0	1.22%	6.55%	33	18	**24.50**	33.61
rd100	7904	7896.5	7334.0	0.09%	7.21%	28	27	**31.29**	41.14
eil101	629	627.5	619.0	0.24%	1.59%	11	13	**14.45**	65.25
lin105	14379	14370.5	13213.5	0.06%	8.11%	40	29	44.46	**29.69**
pr107	44303	44303.0	43381.0	0.00%	2.08%	168	35	173.08	**10.50**
pr124	59030	58067.5	50200.0	1.63%	14.96%	76	23	88.97	**47.34**
bier127	118282	117431.0	112278.5	0.72%	5.08%	40	28	**62.66**	146.80
ch130	6110	6075.5	5597.0	0.56%	8.40%	50	35	**85.77**	143.49
pr144	58537	58189.3	32863.0	0.59%	43.86%	59	69	234.70	**192.84**
ch150	6528	6490.1	6295.0	0.58%	3.57%	17	32	**87.24**	477.22
kroA150	26524	26299.0	24848.8	0.85%	6.32%	30	47	**127.02**	733.44
kroB150	26130	25732.5	24698.0	1.52%	5.48%	53	30	**116.62**	358.89
u159	42080	41925.0	40685.0	0.37%	3.32%	21	29	**85.87**	1183.81
si175	21407	21374.5	21140.0	0.15%	1.25%	84	40	**287.76**	799.61
brg180	1950	1950.0	1800.0	0.00%	7.69%	29	17	**113.69**	2671.59

Fig. 8.6 Table 6.1 from Ardekani [90]

Table 6.2: Results for the LP Relaxation of MI Formulation for Some TSPLIB Instances

Problem	Size	Optimal Value	MI relaxation	MI Gap	Cplex Iterations	Solution Seconds
gr17	17	2085	2085.0	0.00%	133	0.02
gr21	21	2707	2707.0	0.00%	298	0.02
gr24	24	1272	1272.0	0.00%	288	0.03
fri26	26	937	937.0	0.00%	224	0.02
bayg29	29	1610	1608.0	0.12%	375	0.03
bays29	29	2020	2013.5	0.32%	435	0.03
swiss42	42	1273	1272.0	0.08%	960	0.14
dantzig42	42	699	697.0	0.29%	935	0.16
att48	48	10628	10604.0	0.23%	2028	0.58
gr48	48	5046	4959.0	1.72%	2363	0.69
hk48	48	11461	11444.5	0.14%	1820	0.52
eil51	51	426	422.5	0.82%	2217	0.88
berlin52	52	7542	7542.0	0.00%	2794	0.97
brazil58	58	25395	25345.5	0.19%	3596	1.42
st70	70	675	671.0	0.59%	4838	3.39
eil76	76	538	537.0	0.19%	7983	9.92
pr76	77	108159	105120.0	2.81%	3380	2.80
rat99	99	1211	1206.0	0.41%	7994	14.25
kroA100	100	21282	20936.5	1.62%	11708	29.89
kroB100	100	22141	21834.0	1.39%	10732	25.25
kroC100	100	20749	20472.5	1.33%	9386	21.88
kroD100	100	21294	21141.5	0.72%	11058	27.78
kroE100	100	22068	21799.5	1.22%	12338	33.61
rd100	100	7904	7896.5	0.09%	12603	41.14
eil101	101	629	627.5	0.24%	17586	65.25
lin105	105	14379	14371.0	0.06%	7358	29.69
pr107	107	44303	44303.0	0.00%	5988	10.50
gr120	120	6942	6911.3	0.44%	27553	151.81
pr124	124	59030	58067.5	1.63%	9792	47.34
bier127	127	118282	117431.0	0.72%	22753	146.80
ch130	130	6110	6075.5	0.56%	24542	143.49
pr136	136	96772	95934.5	0.87%	31818	216.08
pr144	144	58537	58189.3	0.59%	19418	192.84
ch150	150	6528	6490.1	0.58%	43906	477.22
kroA150	150	26524	26299.0	0.85%	49837	733.44
kroB150	150	26130	25732.5	1.52%	35071	358.89
pr152	152	73682	73208.5	0.64%	24364	224.98
u159	159	42080	41925.0	0.37%	80621	1183.81
si175	175	21407	21374.5	0.15%	47553	799.61
brg180	180	1950	1950.0	0.00%	135584	2671.59
rat195	195	2323	2299.3	1.02%	52534	706.19
d198	198	15780	15722.0	0.37%	69002	958.75
kroA200	200	29368	29065.0	1.03%	93262	3154.19
kroB200	200	29437	28865.0	1.94%	99720	3043.31

Continued on next page

Fig. 8.7   Table 6.2 from Ardekani [90]

Problem	Size	Optimal Value	MI relaxation	MI Gap	Cplex Iterations	Solution Seconds
pr226	226	80369	80092.0	0.34%	141296	6776.33
pr264	264	49135	49020.5	0.23%	90859	2022.39
a280	280	2579	2566.0	0.50%	412138	34954.51
pr299	299	48191	47423.0	1.59%	439825	35243.33
lin318	318	42029	41888.8	0.33%	289438	23196.38
linhp318	318	41345	41245.8	0.24%	289438	22386.91

Fig. 8.8 Continuation of Table 6.2 from Ardekani [90]

Table 6.15: Cplex Iterations for Solving Directed Diamond with ATSP Formulations

k	Size	Carr	Claus	FGG	Flood	MI	MTZ
2	12	146	1230	10	9	7	10
3	18	1351	5636	16	105	38	16
4	24	3000	16510	24	206	88	22
5	30	5261	34980	30	384	243	28
6	36	12422	66997	37	587	464	35
7	42	24594	120050	52	400	339	42
8	48	43944	191265	58	794	751	46
9	54	58468	296440	69	974	1048	52
10	60	93089	422779	77	1565	1270	58
11	66	133179	633209	78	2256	1599	64
12	72	178982	949096	95	3276	2162	70

Fig. 8.9 Table 6.15 from Ardekani [90]

to the LP relaxation of the *MI*-formulation are equal to the results of the LP relaxation of the *DFJ* formulation. Problems larger than 318 for *MI*-relaxation could not be solved as it exceeded the Cplex memory limit.

Ardekani solved diamond instances [139] from size 16 to 104 using LP relaxations of various *TSP* formulations. We did not solve instances larger than 11-diamonds by Wong [165] formulation, or instances larger than 12-diamonds by Claus [46] formulation, or instances larger than 13-diamonds by Carr [40] formulation as it would exceed the Cplex memory limit. The results of this experiment are presented in Table 6.15 of [90] (reproduced here as Fig. 8.9).

8.8.3 Appendix 3: On the Specialised Version of the Procedure FLOW

Here we give the specialised version of the procedure $FLOW$ from *Minimum Cost Flows in Hypergraphs* by Cambini et al. [38].

This version uses the fact that the tail of a hyperarc in *MI*-relaxation problem is either $\emptyset$ or $\{(i, k), (j, k), k\}$ when the head is (i, j).

Algorithm 5 Procedure flow:

INPUT: Hypergraph $\mathcal{H} = (\mathcal{V}, \mathcal{E})$, a spanning tree given by $T_R = (R \cup N, E_T)$;
a traverse of the spanning tree $T_R = (R, e_1, v_1, \ldots, e_q, v_q)$; demands at nonroot vertices $d(N)$,
flows on external hyperarcs; external hyperarcs $E_X = \mathcal{E} \setminus E_T$.
OUTPUT: demands at root vertices $d(R)$, and flows on tree hyperarcs $f(T)$;

---

for $v \in R$ **do**
  $d(v) = 0$;
end for
for $e = (k : (i, j)) \in E_X$ **do**
  $d(i, k) \leftarrow d(i, k) + f(e)$;
  $d(j, k) \leftarrow d(j, k) + f(e)$;
  $d(k) \leftarrow d(k) + f(e)$;
  $d(i, j) \leftarrow d(i, j) - f(e)$;
end for
for $e = (\emptyset, (i, j)) \in E_X$ **do**
  $d(i, j) \leftarrow d(i, j) - f(e)$;
end for
for $v \in \mathcal{V}$ **do**
  $unvisited(v) \leftarrow$ number of hyperarcs incident into v;
end for
$Queue \leftarrow \{v \mid v$ is a leaf of $T_R\}$;
while $Queue \neq \emptyset$ **do**
  Select a $v \in Queue$;
  $Queue \leftarrow Queue \setminus v$;
  let $e_v = (k' : (i', j'))$;
  $f(e_v) \leftarrow d(v)$ if $v = i'j'$, $-d(v)$ otherwise;
  for $w \in \{i'j', i'k', j'k', k'\} \setminus \{v\}$ **do**
    $d(w) \leftarrow d(w) - f(e_v)$ if $w = i'j'$, $d(w) + f(e_v)$, otherwise;
    $unvisited(w) \leftarrow unvisited(w) - 1$;
    if $unvisited(w) = 1$ and $w \notin R$ **then**
      $Queue \leftarrow Queue \cup \{w\}$;
    end if
  end for
end while
for $v \in R$ **do**
  $d(v) \leftarrow -d(v)$;
end for
return demands at root vertices: $d(R)$, and flows on tree hyperarcs: $f(T)$;

Bibliographical Notes

Here I provide bibliographical notes clarifying the use of the material in different chapters, from previously published journal articles or book chapters by me (Arthanari, with or without co-author(s)) [6–8, 10–19].

Chapter 1

This chapter is new, written by me, and has not been published before. This chapter contains several quotations from articles by many stalwarts in the field. All inserted quotes are fully referenced and attributed.

Chapter 2

Since this chapter is preliminary material, I have gathered many concepts and definitions from different fields, namely, Linear Programming, Graph theory, Computational Complexity, Convex Polytopes and Combinatorial optimisation. References are provided to sources for further input on the topic. Some of the material like that on *FAT* problems, *FFF* algorithm by Gusfield [89] and the related graph theoretic concepts have appeared earlier in my papers [14, 18].

T. S. Arthanari, *Pedigree Polytopes*,
https://doi.org/10.1007/978-981-19-9952-9

Chapter 3

The multistage insertion formulation and some results on the dynamic programming structure were earlier obtained by me in 1982. *MI*-formulation first appears in [10], see the abstract in [20]. The material in this chapter, especially the connection between *MI*-relaxation and subtour elimination polytope was first shown in the paper co-authored with Usha Mohan [18]. The comment on Petersen's graph by Donald Knuth is from: The art of computer programming, Volume 3, Sorting and Searching [105].

Chapter 4

A major part of this chapter comes from my journal article published in *Discrete Mathematics*, entitled'On Pedigree polytope and Hamiltonian cycles' [14]. A sufficiency condition for nonadjacency in the tour polytope presented in Sect. 4.6 is from my journal article published in *Discrete Optimization*. While discussing the diameter of the pedigree polytope in Sect. 4.6.2 the material on the asymptotic behaviour of the diameter of the pedigree polytope, I have paraphrased, keeping the pedigree framework intact, from the work of Makkeh et al. [116, 117], who use the corresponding tours in their discussions. The quote on 'the possibility of there may exist "good" algorithms for a large class of problems' is from the journal article by Padberg & Rao entitled, 'The Travelling salesman problem and a class of polyhedra of diameter two', published in *Mathematical Programming*. After four years, in 1978, Papadimitriou published his important result that the problem of nonadjacency testing for tour polytope is in NP-complete class, as a way to express his disbelief in Padberg and Rao's expectation. Much later Matsui [119] showed many more polytopes have the same fate as the tour-polytope, concerning testing nonadjacency in their 1-skeleton or graph of the polytope. Recently, Maksimenko [118] showed one of those polytopes (double covering polytopes) studied by Matsui, as the common face of some 0/1-polytopes for which the non-adjacency testing problem is NP-complete. It will be of interest how this observation could be useful in devising polynomial nonadjacency testing algorithms for 0/1 polytopes, given the major result from Chap. 7.

Chapter 5

I have presented my research in this chapter that has not been published elsewhere. However, earlier a necessary condition for membership in the pedigree polytope was presented by me in Chapter 6 of the book *Mathematical Programming and Game Theory for Decision Making*, pp. 61–98, published by World Scientific.

A necessary and sufficient condition presented in this chapter (of the present book) and its validation are new. In addition, the restricted network defined here has additional requirements not stated in the aforementioned book chapter [12]. Moreover, the layered networks (N_k, R_k, μ) defined here are new. Thus the results and proofs though they are similar to that in [12], are rewritten to incorporate and check the new requirements, hence the need to refer to the earlier work is avoided.

Chapter 6

This chapter emanates from my new research and is not published elsewhere. The results on the mutual adjacency of the pedigrees in any set of rigid pedigrees are new. These are crucial for showing the framework, given to check membership in pedigree polytope, is implementable in strongly polynomial time.

The example given in Sect. 6.3 was used in the paper (with Laleh Ardekani) [7] to show that the necessary condition given in [12] is not sufficient. However, the use of the example to illustrate and run through the steps of the framework and check the necessary and sufficient condition is new.

Chapter 7

The technicalities that need to be checked to go from membership oracle/subroutine for a polytope to the conclusion that linear optimisation over that polytope can be solved efficiently were already known from the book *Geometric Algorithms and Combinatorial Optimization*. Professors Martin Grötschel, László Lovász, and Alexander Schrijver have done a great service to the optimisation community through this work. I need to mention here, while I was working on some separation results for pedigree polytope using polyhedral separating inequalities, instead of linear separation, I wrote an e-mail to Professor Lovász for his opinion. And his suggestion that I could work on the membership problem instead was crucial to divert my attention to the membership problem studied which resulted in [12]. I have taken some results from my earlier work in this chapter (presented in Sects. 7.2 and 7.3) for validating that the pedigree optimisation problem is polynomially solvable. Thus showing *STSP* problem can be solved efficiently is an unexpected new result, a major contribution from this book.

The Appendix on Maurras' construction is excerpted with permission from the article by Maurras [121], entitled, 'From membership to separation, a simple construction', published in *Combinatorica*, volume 22, issue 4, 531–536. Here we also refer to his later work [122].

Chapter 8

This chapter brings together some results of the computational experiments done with *MI*-formulation earlier and some new algorithmic possibilities arising from the pedigree approach to solving *STSP* reported by me in [15, 17].

1 Computational experiments were performed that compared different *TSP* formulations with the *MI*-formulation.

    (a) by Dr Laleh Haerian Ardekani as part of her doctoral project at the University of Auckland [90], entitled, 'New Insights on the Multistage Insertion Formulation of the Traveling Salesman Problem Polytopes, Algorithms, and Experiments' and

    (b) by Mr M Gubb as part of his project at the Department of Engineering Science and Biomedical Engineering, University of Auckland [86].

Some of these computational comparison results presented in Sect. 8.2 have appeared earlier in my article entitled 'On pedigree polytope and its properties' [15].

2 Sections on *MI*-formulation and hypergraph flows, and the Lagrangian approach, have appeared earlier in my article entitled, 'Symmetric traveling salesman problem and flows in hypergraphs: New algorithmic possibilities' [15]. And the plan of the computational experiments and material in Sect. 8.4 have appeared in [13] our joint work (with Kun Qian), a chapter in *Mathematical Programming and Game Theory*, pp. 87–114, published by Springer, 2018.

3 Sections 8.5 and 8.6 and the appendix on Sect. 8.8.1 are based on our joint conference paper (with Dr Laleh Ardekani and Professor Matthias Ehrgott), entitled, 'Performance of the branch and bound algorithm on the multistage insertion formulation of the traveling salesman problem', in Proceedings of the 45th Annual Conference of the ORSNZ.

References

1. Ahuja, R. K., Magnanti, T. L., & Orlin, J. B. (1988). *Network flows*. Cambridge, Massachusetts: Alfred P. Sloan School of Management.
2. Ahuja, R. K., Magnanti, T. L., & Orlin, J. B. (1996). *Network flows theory, algorithms and applications*. Englewood Cliffs, NJ: Prentice Hall.
3. Anstreicher, K. M. (1997). On Vaidya's volumetric cutting plane method for convex programming. *Mathematics of Operations Research, 22*(1), 63–89.
4. Applegate, D. L., Bixby, R. E., Chvátal, V., Cook, W., Espinoza, D. G., Goycoolea, M., & Helsgaun, K. (2009). Certification of an optimal TSP tour through 85,900 cities. *Operations Research Letters, 37*(1), 11–15.
5. Applegate, D. L., Bixby, R. E., Chvátal, V., & Cook, W. J. (2011). *The traveling salesman problem*. Princeton: Princeton University Press.
6. Ardekani, L. H., & Arthanari, T. S. (2008). Traveling salesman problem and membership in pedigree polytope-a numerical illustration. In *International Conference on Modelling, Computation and Optimization in Information Systems and Management Sciences* (pp. 145–154). Berlin: Springer.
7. Ardekani, L. H., & Arthanari, T. S. (2013). The multi-flow necessary condition for membership in the pedigree polytope is not sufficient- a counterexample. In *Proceedings of the 5th Asian Conference on Intelligent Information and Database Systems - Volume Part II*, ACIIDS'13 (pp. 409–419). Berlin, Heidelberg. Springer.
8. Ardekani, L. H., Arthanari, T. S., & Ehrgott, M. (2010). Performance of the branch and bound algorithm on the multistage insertion formulation of the traveling salesman problem. In *Proceedings of the 45th Annual Conference of the ORSNZ* (pp. 326–335).
9. Arthanari, T. (1974). *On some problems of sequencing and grouping*. Ph.D. thesis, Indian Statistical Institute-Kolkata.
10. Arthanari, T. (1982). On the traveling salesman problem. In *XI Symposium on Mathematical Programming held at Bonn, West Germany*.
11. Arthanari, T. (2005). Pedigree polytope is a combinatorial polytope. In *Operations research with economic and industrial applications: Emerging trends* (pp. 1–17). Delhi: Anamaya Publishers.
12. Arthanari, T. (2008). On the membership problem of the pedigree polytope. In *Mathematical programming and game theory for decision making* (pp. 61–98). Singapore: World Scientific.
13. Arthanari, T., & Qian, K. (2018). Symmetric travelling salesman problem. In *Mathematical programming and game theory* (pp. 87–114). Berlin: Springer.

14. Arthanari, T. S. (2006). On pedigree polytopes and Hamiltonian cycles. *Discrete Mathematics, 306*(14), 1474–1492.
15. Arthanari, T. S. (2013). On pedigree polytope and its properties. *Atti della Accademia Peloritana dei Pericolanti-Classe di Scienze Fisiche, Matematiche e Naturali, 91*(S2).
16. Arthanari, T. S. (2013). Study of the pedigree polytope and a sufficiency condition for non-adjacency in the tour polytope. *Discrete Optimization, 10*(3), 224–232.
17. Arthanari, T. S. (2019). Symmetric traveling salesman problem and flows in hypergraphs: New algorithmic possibilities. *Atti della Accademia Peloritana dei Pericolanti-Classe di Scienze Fisiche, Matematiche e Naturali, 97*(1), 1.
18. Arthanari, T. S., & Usha, M. (2000). An alternate formulation of the symmetric traveling salesman problem and its properties. *Discrete Applied Mathematics, 98*(3), 173–190.
19. Arthanari, T. S., & Usha, M. (2001). On the equivalence of the multistage-insertion and cycle-shrink formulations of the symmetric traveling salesman problem. *Operations Research Letters, 29*(3), 129–139.
20. Bachem, A., Grötschel, M., & Korte, B. (2012). *Mathematical programming the state of the art: Bonn 1982*. Berlin: Springer Science & Business Media.
21. Balas, E., & Padberg, M. (1975). On the set-covering problem: II. An algorithm for set partitioning. *Operations Research, 23*(1), 74–90.
22. Balas, E., & Padberg, M. W. (1972). On the set-covering problem. *Operations Research, 20*(6), 1152–1161.
23. Balas, E., & Padberg, M. W. (1976). Set partitioning: A survey. *SIAM Review, 18*(4), 710–760.
24. Balas, E., & Toth, P., et al. (1985). Branch and bound methods. In E. L. Lawler (Ed.), *The traveling salesman problem: A guided tour of combinatorial optimization* (pp. 107–132). New York: Wiley.
25. Balinski, M., & Russakoff, A. (1974). On the assignment polytope. *Siam Review, 16*(4), 516–525.
26. Barahona, F., & Anbil, R. (2000). The volume algorithm: Producing primal solutions with a subgradient method. *Mathematical Programming, 87*(3), 385–399.
27. Barnette, D. (1974). An upper bound for the diameter of a polytope. *Discrete Mathematics, 10*(1), 9–13.
28. Bellman, R. (1962). Dynamic programming treatment of the travelling salesman problem. *Journal of the ACM (JACM), 9*(1), 61–63.
29. Bellmore, M., & Malone, J. C. (1971). Pathology of traveling-salesman subtour-elimination algorithms. *Operations Research, 19*(2), 278–307.
30. Berge, C. (1957). Two theorems in graph theory. *Proceedings of the National Academy of Sciences of the United States of America, 43*(9), 842–844.
31. Bertsimas, D., & Orlin, J. B. (1994). A technique for speeding up the solution of the lagrangean dual. *Mathematical Programming, 63*(1–3), 23–45.
32. Billera, L. J. (1983). Polyhedral theory and commutative algebra. In *Mathematical programming the state of the art* (pp. 57–77). Berlin: Springer.
33. Billera, L. J., & Sarangarajan, A. (1996). All 0–1 polytopes are traveling salesman polytopes. *Combinatorica, 16*(2), 175–188.
34. Black, A., De Loera, J., Kafer, S., & Sanità, L. (2021). On the simplex method for 0/1 polytopes. arXiv:2111.14050.
35. Bland, R. G., Goldfarb, D., & Todd, M. J. (1981). The ellipsoid method: A survey. *Operations Research, 29*(6), 1039–1091.
36. Bollobás, B. (2013). *Modern graph theory* (vol. 184). Berlin: Springer Science & Business Media.
37. Brondsted, A. (2012). *An introduction to convex polytopes* (vol. 90). Berlin: Springer Science & Business Media.
38. Cambini, R., Gallo, G., & Scutellà, M. (1992). *Minimum Cost Flows on Hypergraphs*. Technical report (Università di Pisa. Dipartimento di informatica). Università degli studi di Pisa, Dipartimento di informatica.

39. Cambini, R., Gallo, G., & Scutellà, M. G. (1997). Flows on hypergraphs. *Mathematical Programming, 78*(2), 195–217.
40. Carr, R. (1996). Separating over classes of TSP inequalities defined by 0 node-lifting in polynomial time. In *International Conference on Integer Programming and Combinatorial Optimization* (pp. 460–474). Berlin: Springer.
41. Carr, R. D., & Konjevod, G. (2005). Polyhedral combinatorics. In *Tutorials on emerging methodologies and applications in Operations Research* (pp. 2–1). Berlin: Springer.
42. Christof, T., & Reinelt, G. (2001). Decomposition and parallelization techniques for enumerating the facets of combinatorial polytopes. *International Journal of Computational Geometry & Applications, 11*(04), 423–437.
43. Christofides, N. (1976). Worst-case analysis of a new heuristic for the travelling salesman problem. Technical report, Carnegie-Mellon Univ Pittsburgh Pa Management Sciences Research Group.
44. Christos, H. P., & Steiglitz, K. (1982). *Combinatorial optimization: Algorithms and complexity*. Prentice: Prentice Hall Inc.
45. Chvátal, V. (1975). On certain polytopes associated with graphs. *Journal of Combinatorial Theory, Series B, 18*(2), 138–154.
46. Claus, A. (1984). A new formulation for the travelling salesman problem. *SIAM Journal on Discrete Mathematics, 5*(1), 21–25.
47. Colman, A. M. (2009). *A dictionary of psychology*. Oxford: Oxford University Press.
48. Cook, S. (2006). The P versus NP problem. In J. A. Carlson, A. Jaffe, & A. Wiles (Eds.), *The millennium prize problems* (pp. 87–104). Providence: American Mathematical Society.
49. Cook, S. A. (1971). The complexity of theorem-proving procedures. In *Proceedings of the third annual ACM symposium on Theory of computing* (pp. 151–158).
50. Cook, W. (2005). Concorde home. http://www.tsp.gatech.edu/concorde.html.
51. Cook, W. (2019). Computing in combinatorial optimization. In *Computing and Software Science* (pp. 27–47). Berlin: Springer.
52. Cook, W. J., Cunningham, W., Pulleyblank, W., & Schrijver, A. (2009). Combinatorial optimization. *Oberwolfach Reports, 5*(4), 2875–2942.
53. Cplex, I. I. (2009). V12. 1: User's manual for cplex. *International Business Machines Corporation, 46*(53), 157.
54. Crowder, H., & Padberg, M. W. (1980). Solving large-scale symmetric travelling salesman problems to optimality. *Management Science, 26*(5), 495–509.
55. Dahl, G. (1997). *An introduction to convexity, polyhedral theory and combinatorial optimization*. University of Oslo, Department of Informatics.
56. Dantzig, G., Fulkerson, R., & Johnson, S. (1954). Solution of a large-scale traveling-salesman problem. *Journal of the Operations Research Society of America, 2*(4), 393–410.
57. Dantzig, G. B. (1951). Application of the simplex method to a transportation problem. *Activity analysis and production and allocation*.
58. Dantzig, G. B. (1983). Reminiscences about the origins of linear programming. In *Mathematical programming the state of the art* (pp. 78–86). Berlin: Springer.
59. Dantzig, G. B. (1990). Origins of the simplex method. In *A history of scientific computing* (pp. 141–151). Association of Computing Machinaries.
60. Dell'Amico, M., Maffioli, F., & Martello, S. (1997). *Annotated bibliographies in combinatorial optimization*. New York: Wiley.
61. Diestel, R. (2017). The basics. In *Graph theory* (pp. 1–34). Berlin: Springer.
62. Eastman, W. (1958). *Linear Programming with Pattern Constraints*. Ph. D. Thesis, Department of Economics, Harvard University, Cambridge, Massachusetts, USA.
63. Edelman, A. (2019). Julia: A fresh approach to technical computing and data processing. Technical report, Massachusetts Institute of Technology, Cambridge.
64. Edmonds, J. (1965). Paths, trees, and flowers. *Canadian Journal of Mathematics, 17*, 449–467.
65. Edmonds, J. (1967). Systems of distinct representatives and linear algebra. *Journal of Research of the National Bureau of Standards, Section B, 71*(4), 241–245.

66. Edmonds, J. and Giles, R. (1977). A min-max relation for submodular functions on graphs. In *Annals of discrete mathematics* (vol. 1, pp. 185–204). Amsterdam: Elsevier.

67. Edmonds, J., & Johnson, E. L. (2003). *Matching: A well-solved class of integer linear programs* (pp. 27–30). Berlin, Heidelberg: Springer.

68. Edmonds, J., & Karp, R. M. (1972). Theoretical improvements in algorithmic efficiency for network flow problems. *Journal of the ACM (JACM), 19*(2), 248–264.

69. Fischetti, M., Lodi, A., & Toth, P. (2003). Solving real-world atsp instances by branch-and-cut. In *Combinatorial Optimization-Eureka, You Shrink!*, pp. 64–77. Berlin: Springer.

70. Fleischmann, B. (1988). A new class of cutting planes for the symmetric travelling salesman problem. *Mathematical Programming, 40*(1), 225–246.

71. Flood, M. M. (1956). The traveling-salesman problem. *Operations Research, 4*(1), 61–75.

72. Ford, L. R., & Fulkerson, D. R. (1957). A simple algorithm for finding maximal network flows and an application to the hitchcock problem. *Canadian Journal of Mathematics, 9*, 210–218.

73. Ford, L. R., & Fulkerson, D. R. (2015). *Flows in networks*. Princeton: Princeton University Press.

74. Fox, K., Gavish, B., & Graves, S. (1980). An n-constraint formulation of the time-dependent travelling salesman problem. *Operations Research, 28*(4), 1018–1021.

75. Fulkerson, D. (1956). *Hitchcock transportation problem*. Technical report, Rand Corp, Santa Monica, CA.

76. Garey, M. R., & Johnson, D. S. (1979). *Computers and intractability* (vol. 174). Freeman San Francisco.

77. Geoffrion, A. M. (1974). Lagrangean relaxation for integer programming. In *Approaches to integer programming* (pp. 82–114). Berlin: Springer.

78. Godinho, M. T., Gouveia, L., & Pesneau, P. (2014). Natural and extended formulations for the time-dependent traveling salesman problem. *Discrete Applied Mathematics, 164*, 138–153.

79. Gomory, R. E. (1958). Outline of an algorithm for integer solutions to linear programs. *Bulletin of the American Mathematical Society, 64*, 275–278.

80. Gouveia, L., & Pires, J. M. (1999). The asymmetric travelling salesman problem and a reformulation of the miller-tucker-zemlin constraints. *European Journal of Operational Research, 112*(1), 134–146.

81. Gouveia, L., & Voß, S. (1995). A classification of formulations for the (time-dependent) traveling salesman problem. *European Journal of Operational Research, 83*(1), 69–82.

82. Grötschel, M., Lovász, L., & Schrijver, A. (2012). *Geometric algorithms and combinatorial optimization* (vol. 2). Berlin: Springer Science & Business Media.

83. Grötschel, M., & Nemhauser, G. L. (2008). George dantzig's contributions to integer programming. *Discrete Optimization, 5*(2), 168–173.

84. Grötschel, M., & Padberg, M. W. (1979). On the symmetric travelling salesman problem I: Inequalities. *Mathematical Programming, 16*(1), 265–280.

85. Grötschel, M., & Padberg, M. W., et al. (1985). Polyhedral theory. In E. L. Lawler (Ed.), *The traveling salesman problem: A guided tour of combinatorial optimization* (pp. 107–132). Berlin: Wiley.

86. Gubb, M. (2003). Flows, insertions and subtours- modelling the travelling salesman. Project report, Part IV project, Engineering Science & Biomedical Engineering, University of Auckland, New Zealand.

87. Guignard, M. (2003). Lagrangean relaxation. *Top, 11*(2), 151–200.

88. Gurobi Optimization, LLC. (2021). Gurobi optimizer reference manual.

89. Gusfield, D. (1988). A graph theoretic approach to statistical data security. *SIAM Journal on Computing, 17*(3), 552–571.

90. Haerian Ardekani, L. (2011). *New Insights on the Multistage Insertion Formulation of the Traveling Salesman Problem-Polytopes, Experiments, and Algorithm*. Ph.D. thesis, Business School, University of Auckland, New Zealand.

91. Held, M., & Karp, R. M. (1962). A dynamic programming approach to sequencing problems. *Journal of the Society for Industrial and Applied mathematics, 10*(1), 196–210.

92. Held, M., & Karp, R. M. (1970). The traveling-salesman problem and minimum spanning trees. *Operations Research, 18*(6), 1138–1162.
93. Held, M., & Karp, R. M. (1971). The traveling-salesman problem and minimum spanning trees: Part ii. *Mathematical Programming, 1*(1), 6–25.
94. Heller, I., et al. (1956). Neighbor relations on the convex of cyclic permutations. *Pacific Journal of Mathematics, 6*(3), 467–477.
95. Houck, D. J., Jr., Picard, J.-C., Queyranne, M., & Vemuganti, R. R. (1978). Traveling salesman problem as a constrained shortest path problem: Theory and computational experience. *Opsearch, 17*, 93–109.
96. (https://mathoverflow.net/users/18060/willsawin), W. S. (2012). H-representation versus V-representation of polytopes. MathOverflow. https://mathoverflow.net/q/108715 (version: 2012-10-03).
97. Ikebe, Y., Matsui, T., & Tamura, A. (1993). Adjacency of the best and second best valued solutions in combinatorial optimization problems. *Discrete Applied Mathematics, 47*(3), 227–232.
98. Ikura, Y., & Nemhauser, G. L. (1985). Simplex pivots on the set packing polytope. *Mathematical Programming, 33*(2), 123–138.
99. Jeroslow, R. G., Martin, K., Rardin, R. L., & Wang, J. (1992). Gainfree Leontief substitution flow problems. *Mathematical Programming, 57*(1), 375–414.
100. Karmarkar, N. (1984). A new polynomial-time algorithm for linear programming. In *Proceedings of the sixteenth annual ACM symposium on Theory of computing* (pp. 302–311).
101. Karp, R. M. (1972). Reducibility among combinatorial problems. In *Complexity of computer computations* (pp. 85–103). Berlin: Springer.
102. Karp, R. M. (1975). On the computational complexity of combinatorial problems. *Networks, 5*(1), 45–68.
103. Khachiyan, L. G. (1979). A polynomial algorithm in linear programming. In *Doklady Akademii Nauk* (vol. 244, 5, pp. 1093–1096). Moscow: Russian Academy of Sciences.
104. Klee, V., & Minty, G. J. (1972). How good is the simplex algorithm. *Inequalities, 3*(3), 159–175.
105. Knuth, D. E. (1997). *The art of computer programming* (vol. 3). London: Pearson Education.
106. Knuth, D. E. (2011). *The art of computer programming, volume 4A: combinatorial algorithms, part 1*. Pearson Education India.
107. Koopmans, T. C., & Reiter, S. (1951). A model of transportation. *Activity analysis of production and allocation* (pp. 222–259).
108. Korte, B. H., & Vygen, J. (2011). *Combinatorial optimization* (vol. 1). Berlin: Springer.
109. Lawler, E. L. (2001). *Combinatorial optimization: networks and matroids*. North Chelmsford: Courier Corporation.
110. Lawler, E. L., & Wood, D. E. (1966). Branch-and-bound methods: A survey. *Operations Research, 14*(4), 699–719.
111. Lemaréchal, C. (2007). The omnipresence of Lagrange. *Annals of Operations Research, 153*(1), 9–27.
112. Letchford, A. N., & Lodi, A. (2010). Mathematical programming approaches to the traveling salesman problem. *Wiley Encyclopedia of Operations Research and Management Science*.
113. Little, J. D., Murty, K. G., Sweeney, D. W., & Karel, C. (1963). An algorithm for the traveling salesman problem. *Operations Research, 11*(6), 972–989.
114. Lorena, L. A. N., & Narciso, M. G. (2002). Using logical surrogate information in Lagrangean relaxation: An application to symmetric traveling salesman problems. *European Journal of Operational Research, 138*(3), 473–483.
115. Lovász, L. (1986). *An algorithmic theory of numbers, graphs and convexity*. SIAM.
116. Makkeh, A., Pourmoradnasseri, M., & Theis, D. O. (2016). On the graph of the pedigree polytope. arXiv:1611.08431.
117. Makkeh, A., Pourmoradnasseri, M., & Theis, D. O. (2017). The graph of the pedigree polytope is asymptotically almost complete. In *Conference on algorithms and discrete applied mathematics* (pp. 294–307). Berlin: Springer.

118. Maksimenko, A. (2014). The common face of some 0/1-polytopes with NP-complete nonadjacency relation. *Journal of Mathematical Sciences, 203*(6), 823–832.
119. Matsui, T. (1995). Np-completeness of non-adjacency relations on some 0-1-polytopes. *Lecture Notes in Operations Research, 1*, 249–258.
120. Matsuia, T., & Tamura, S. (1995). Adjacency on combinatorial polyhedra. *Discrete Applied Mathematics, 56*(2–3), 311–321.
121. Maurras, J. F. (2002). From membership to separation, a simple construction. *Combinatorica, 22*(4), 531–536.
122. Maurras, J. F. (2010). Note on separation from membership, and folklore. *Mathematical Programming, 124*(1–2), 7.
123. Miller, C., Tucker, A., & Zemlin, R. (1960). Integer programming formulations and traveling salesman problems. *Journal of the Association for Computing Machinery, 7*(4), 326–329.
124. Munkres, J. R. (2018). *Elements of algebraic topology*. Boca Raton: CRC Press.
125. Murthy, U., & Bondy, J. (1976). *Graph theory with applications*. New York: The Macmillan Press Ltd.
126. Murty, K. G. (1969). On the tours of a traveling salesman. *SIAM Journal on Control, 7*(1), 122–131.
127. Murty, K. G., & Yu, F.-T. (1988). *Linear complementarity, linear and nonlinear programming* (vol. 3). Berlin: Heldermann.
128. Naddef, D. (1989). The Hirsch conjecture is true for (0, 1)-polytopes. *Mathematical Programming: Series A and B, 45*(1–3), 109–110.
129. Naddef, D. (2007). Polyhedral theory and branch-and-cut algorithms for the symmetric TSP. In *The traveling salesman problem and its variations* (pp. 29–116). Berlin: Springer.
130. Naddef, D., & Pulleyblank, W. R. (1981). Hamiltonicity and combinatorial polyhedra. *Journal of Combinatorial Theory, Series B, 31*(3), 297–312.
131. Nedić, A., & Ozdaglar, A. (2009). Approximate primal solutions and rate analysis for dual subgradient methods. *SIAM Journal on Optimization, 19*(4), 1757–1780.
132. Öncan, T., Altınel, I. K., & Laporte, G. (2009). A comparative analysis of several asymmetric traveling salesman problem formulations. *Computers & Operations Research, 36*(3), 637–654.
133. Orlin, J. B. (1985). On the simplex algorithm for networks and generalized networks. In R. W. Cottle (Ed.), *Mathematical Programming Essays in Honor of George B. Dantzig Part I* (pp. 166–178). Berlin, Heidelberg: Springer.
134. Orman, A., & Williams, H. P. (2007). A survey of different integer programming formulations of the travelling salesman problem. In *Optimisation, econometric and financial analysis* (pp. 91–104). Berlin: Springer.
135. Padberg, M. (2013). *Linear optimization and extensions* (vol. 12). Berlin: Springer Science & Business Media.
136. Padberg, M., & Sung, T.-Y. (1991). An analytical comparison of different formulations of the travelling salesman problem. *Mathematical Programming, 52*(1), 315–357.
137. Padberg, M. W., & Rao, M. R. (1974). The travelling salesman problem and a class of polyhedra of diameter two. *Mathematical Programming, 7*(1), 32–45.
138. Papadimitriou, C. H. (1978). The adjacency relation on the traveling salesman polytope is NP-complete. *Mathematical Programming, 14*(1), 312–324.
139. Papadimitriou, C. H., & Steiglitz, K. (1977). On the complexity of local search for the traveling salesman problem. *SIAM Journal on Computing, 6*(1), 76–83.
140. Polyak, B. T. (1967). A general method for solving extremal problems. In *Doklady Akademii Nauk* (vol. 174, pp. 33–36). Moscow: Russian Academy of Sciences.
141. Polyak, B. T. (1969). Minimization of unsmooth functionals. *USSR Computational Mathematics and Mathematical Physics, 9*(3), 14–29.
142. Pourmoradnasseri, M. (2017). *Some problems related to extensions of polytopes*. Ph.D. thesis, Tartu University.
143. Pulleyblank, W. R. (1983). Polyhedral combinatorics. In *Mathematical Programming The State of the Art* (pp. 312–345). Berlin: Springer.

144. Pulleyblank, W. R. (2012). Edmonds, matching and the birth of polyhedral combinatorics. *Documenta Mathematica*, 1.
145. Rao, M. (1976). Adjacency of the traveling salesman tours and 0–1 vertices. *SIAM Journal on Applied Mathematics, 30*(2), 191–198.
146. Reinelt, G. (1991). TSPLIB– a traveling salesman problem library. *ORSA Journal on Computing, 3*(4), 376–384.
147. Reinelt, G. (1991). TSPLIB-a traveling salesman problem library. *ORSA Journal on Computing, 3*(4), 376–384.
148. Reinelt, G. (2014). {TSPLIB}: A library of sample instances for the TSP (and related problems) from various sources and of various types. http://comopt.ifi.uniheidelberg.de/software/TSPLIB95.
149. Rispoli, F. J., & Cosares, S. (1998). A bound of 4 for the diameter of the symmetric traveling salesman polytope. *SIAM Journal on Discrete Mathematics, 11*(3), 373–380.
150. Roberti, R., & Toth, P. (2012). Models and algorithms for the asymmetric traveling salesman problem: An experimental comparison. *EURO Journal on Transportation and Logistics, 1*(1–2), 113–133.
151. Rockafellar, R. T. (1970). *Convex analysis* (vol. 18). Princeton: Princeton University Press.
152. Schrijver, A. (1986). *Theory of linear and integer programming*. Chichester: Wiley.
153. Schrijver, A. (2002). On the history of the transportation and maximum flow problems. *Mathematical Programming, 91*(3), 437–445.
154. Schrijver, A. (2003). *Combinatorial optimization: Polyhedra and efficiency* (vol. 24). Berlin: Springer Science & Business Media.
155. Schwinn, J., & Werner, R. (2019). On the effectiveness of primal and dual heuristics for the transportation problem. *IMA Journal of Management Mathematics, 30*(3), 281–303.
156. Sherali, H. D., Sarin, S. C., & Tsai, P.-F. (2006). A class of lifted path and flow-based formulations for the asymmetric traveling salesman problem with and without precedence constraints. *Discrete Optimization, 3*(1), 20–32.
157. Shoemaker, A., & Vare, S. (2016). Edmonds' blossom algorithm. *CME*.
158. Shor, N. Z. (1970). Convergence rate of the gradient descent method with dilatation of the space. *Cybernetics, 6*(2), 102–108.
159. Tardos, É. (1986). A strongly polynomial algorithm to solve combinatorial linear programs. *Operations Research, 34*(2), 250–256.
160. Vaidya, P. M. (1990). An algorithm for linear programming which requires $O(((m+n)n^2 + (m+n)^{1.5}n)L)$ arithmetic operations. *Mathematical Programming, 47*(1), 175–201.
161. Valenzuela, C. L., & Jones, A. J. (1997). Estimating the Held-Karp lower bound for the geometric tsp. *European Journal of Operational Research, 102*(1), 157–175.
162. Veinott, A. F., Jr. (1968). Extreme points of Leontief substitution systems. *Linear Algebra and its Applications, 1*(2), 181–194.
163. Williamson, D. P. (2019). *Network flow algorithms*. Cambridge: Cambridge University Press.
164. Woeginger, G. J. (2003). Exact algorithms for NP-hard problems: A survey. In *Combinatorial optimization-eureka, you shrink!* (pp. 185–207). Berlin: Springer.
165. Wong, R. T. (1980). Integer programming formulations of the traveling salesman problem. In *Proceedings of the IEEE International Conference of Circuits and Computers* (pp. 149–152). IEEE Press Piscataway NJ.
166. Yudin, D., & Nemirovskii, A. S. (1976). Informational complexity and efficient methods for the solution of convex extremal problems. *Matekon, 13*(2), 22–45.
167. Zamani, R., & Lau, S. K. (2010). Embedding learning capability in Lagrangean relaxation: An application to the travelling salesman problem. *European Journal of Operational Research, 201*(1), 82–88.
168. Ziegler, G. M. (2000). Lectures on 0/1-polytopes. In *Polytopes-combinatorics and computation* (pp. 1–41). Berlin: Springer.

Index

Printed in the United States
by Baker & Taylor Publisher Services